A Volume in The Laboratory Animal Pocket Reference Series

Critical Care Management for Laboratory MICE and RATS

F. Claire Hankenson

CRC Press is an imprint of the
Taylor & Francis Group, an **informa** business

CRC Press
Taylor & Francis Group
6000 Broken Sound Parkway NW, Suite 300
Boca Raton, FL 33487-2742

Printed on acid-free paper
Version Date: 20130412

International Standard Book Number-13: 978-0-8493-2499-4 (Paperback)

Library of Congress Cataloging-in-Publication Data

Hankenson, F. Claire, 1971-
Critical care management for laboratory mice and rats / F. Claire Hankenson.
p. ; cm. -- (Laboratory animal pocket reference series)
Includes bibliographical references and index.
ISBN 978-0-8493-2499-4 (pbk.)
I. Title. II. Series: Laboratory animal pocket reference series.
[DNLM: 1. Animals, Laboratory--Handbooks. 2. Laboratory Animal Science--methods--Handbooks. 3. Mice--Handbooks. 4. Rats--Handbooks. QY 39]

SF406
636.088'5--dc23 2013013946

Visit the Taylor & Francis Web site at
http://www.taylorandfrancis.com

and the CRC Press Web site at
http://www.crcpress.com

For Kurt and Shug, who provide me with critical care, supplemental warmth, and essential support (dosed daily, ad libitum)

contents

preface

Progressive scientific and medical advances rely on appropriate selection and study of animal models. In contemporary biomedical research facilities, almost without exception, laboratory mice and rats are examined for optimum health status prior to acceptance as research models. Therefore, common ailments (i.e., respiratory tract infections, diarrhea, and overt skin disease) are often grounds to refuse admittance into an existing animal research colony. If laboratory rodents develop these ailments after acceptance into an experimental study, they most likely will be treated by extrapolation of applicable clinical regimens employed in small animal and exotic pet veterinary practice. Unfortunately, for more complex cases that need critical care management, many treatment approaches to veterinary emergencies cannot be applied directly to the laboratory rodent.

For critical care of laboratory rodents, there is a scarcity of sources that provide comprehensive, feasible, and response-oriented information about clinical interventions specific to spontaneous and induced models of disease. This textbook is the first of its kind devoted to the challenges of critical care management for laboratory mice and rats. This text was designed to serve as a specialized resource for all persons and professionals who utilize rodent models for biomedical research. The information herein emphasizes the varied approaches to laboratory rodent patient care, health assessments, characteristics of specific disease models, monitoring and scoring of disease parameters, and humane interventions. Although preservation of animal health and welfare is of primary consideration, achievement of proposed research objectives should also be prioritized.

Within the field of laboratory animal practice, guidance and recommendations for clinical treatments are rarely uniform across the entire profession. The majority of the dosages and treatments in this book are derived from published abstracts presented at national American Association for Laboratory Animal Science (AALAS) conferences (dating from the year 2000 to 2012), journal articles, and personal consultation and experience. Many treatment modalities are based on anecdotal information, pilot work, and empirical evidence. In the spirit of incorporating continued refinements into laboratory animal practice, these treatment guidelines may be modified further for additional clinical support and improved patient-based care for research animals.

This textbook is organized as five chapters: "General Approaches for Critical Care" (Chapter 1), "Critical Care Management for Laboratory Mice" (Chapter 2), "Critical Care Management for Laboratory Rats" (Chapter 3), "Special Considerations for Critical Care Management in Laboratory Rodents" (Chapter 4), and "Resources and Additional Information" (Chapter 5). In addition, three appendices include a glossary, medical supply list, and an abridged rodent formulary. As endorsed by the *Guide for the Care and Use of Laboratory Animals* (National Research Council, 2011), the expectation is for those clinical recommendations herein to be applied with apt professional judgment, including review of evidence- and performance-based outcomes, while advancing animal welfare and science.

F. Claire Hankenson, DVM, MS, DACLAM

acknowledgments

I would like to thank so many of my colleagues in comparative medicine, all of whom challenge the status quo to identify refinements in laboratory animal practice on a daily basis. The following individuals participated in discussions and information and photograph exchanges during the course of this project: Drs. Valerie Bergdall, Iris Bolton, Gillian Braden, Angela Brice, Ralph Bunte, Andrew Burich, Adam Caro, Anthony Carty, Scout Chou, Laura Conour, Dawn Dinger, Ramon Duran-Struuck, Nancy Figler, Trish Foley, Margaret Fordham, Joanna Fried, Diane Gaertner, Travis Hagedorn, Lisa Halliday, Troy Hallman, Lori Hill, Michael Huerkamp, Marc Hulin, Samer Jaber, Bambi Jasmin, JanLee Jensen, Kari Koszdin, Paul Makidon, Jim Marx, Tom Meier, Angela Mexas, Emily Miedel, Guy Mulder, Stu Leland, John Long, Judy Nielson, Jane Olin, Kate Pritchett-Corning, Jennifer Pullium, Jamie Rhodes, Karen Rosenthal, George Sanders, Kim Saunders, J. Mat Schech, Pat Sharp, Abigail Smith, Jennifer Smith, Peter Smith, Laike Stewart, Doug Taylor, Christin Veeder, Susan Volk, Ashley Wathen, Wendy O. Williams, Jolaine Wilson, and Norm Wiltshire.

My sincere appreciation extends to the staff at the national American Association for Laboratory Animal Science (AALAS) office, particularly John Farrar for his ready assistance with providing materials presented throughout the text. Kimberly Bowe assisted with gathering relevant resource materials, Karena Thek contributed background material for rodent interventions, and Carrie Childs Maute provided advice and comic relief, for which I am indebted. And finally, this work would not have been completed without Dr. Mark Suckow and John Sulzycki and their collective insights, encouragement, flexibility, and patience.

about the author

F. Claire Hankenson, DVM, MS, is the senior associate director in University Laboratory Animal Resources, University of Pennsylvania, Philadelphia, and is an associate professor of laboratory animal medicine in the Department of Pathobiology at the School of Veterinary Medicine. Dr. Hankenson obtained her veterinary degree from Purdue University. Following veterinary school, she completed her laboratory animal medicine residency and graduate work (MS, microbiology) at the University of Washington, Seattle. She became a Diplomate of the American College of Laboratory Animal Medicine (ACLAM) in 2002. Following several years on faculty at the University of Michigan in the Unit for Laboratory Animal Medicine, she transitioned to the University of Pennsylvania. Dr. Hankenson's current position combines administrative service, clinical effort, teaching duties, and collaborative research. Her own research studies involve investigations of refinements in the care and use of laboratory rodents, particularly blood sampling, tail biopsy evaluations, and humane endpoints. Dr. Hankenson has been active on committees in the American Association for Laboratory Animal Science (AALAS) since 2002, has served on the board of directors for ACLAM, and is an ad hoc consultant to AAALAC, International.

general approaches for critical care

overview

The characterization of animal models is becoming increasingly sophisticated. Animal species that reliably mimic human disease provide critical insights so that causative mechanisms can be understood and lead to the development of novel drugs, diagnostic procedures, and therapies. In the ardent hope to advance the clinical practice of contemporary rodent medicine, in housing facilities largely free of infectious diseases, individualized patient care is often prioritized over traditional herd (colony) health diagnostic approaches. As a subsequent benefit, improvements in individual animal health will augment overall colony health measures. Within these colonies, breeding success, production of offspring, and prolonged good health are invaluable. The continued commitment to financing biomedical research and maintaining vast colonies of laboratory rodent models is due to the sheer variety and prevalence of those with unique and irreplaceable genetic backgrounds.

Genetic engineering of laboratory animals is fueled by the desire to unravel the mystery of disease based on contributions by single and multiple genes, molecules and events associated with physiology, development, and function. This dynamic scientific area holds promise for development of new mouse and rat strains that rapidly contribute to and further define applications of in vivo models for research programs (Croy et al., 2001). Biomedical researchers should be aware that both inherent and induced mutations may result in unexpected phenotypes and disease syndromes; thus, genotype and secondary influences on the functionality of the animal's immune system must always be considered, along with housing, husbandry, experimental treatments, age, gender, and body condition, when formulating clinical treatment plans. Increased susceptibility of laboratory rodents to environmental variances, nosocomial bacteria, and infectious diseases will ultimately require heightened veterinary care. Oversight by veterinary specialists, in collaboration with scientific investigators, should ensure that all procedures for disease prevention, diagnosis, and therapy are those currently accepted in veterinary and laboratory animal practice (National Research Council, 2011).

Assessment of clinical abnormalities in rodents may be difficult and complex due to the stoic nature of these species. Exposures to stressors (environmental or otherwise) should be kept to a minimum; rodents have an innate response to escape from a perceived "predator," which can result in harm to themselves and a release of systemic catecholamines that can predispose these species to respiratory and cardiac arrest and severe pulmonary, cardiac, and renal hypertension (Morrisey, 2003). Therefore, if the ultimate objective is to maintain the highest standard of animal welfare, then outcomes and treatments that maintain physiologic stability are ideal.

In rodents that are predisposed highly to stress, the rapidity of the assessment and subsequent patient stabilization are essential prior to complete evaluation for definitive diagnoses of clinical abnormalities (Hawkins and Graham, 2007). Supporting a patient ("supportive care") on initial presentation can be life saving, particularly with the prompt provision of oxygen, fluid therapy, and warmth (Doneley, 2005). Murine intensive care units (M-ICUs), akin to an ICU in veterinary clinical practice, are extremely rare to nonexistent in the biomedical research environment; therefore, treatment plans for critical case management must realistically factor in the unavailability of 24-h healthcare monitoring capabilities.

Undesirable infectious pathogens and parasites that enter facilities and affect rodent colony health should *always* be included in lists of diagnostic differentials, but details on treatment and eradication of these specific agents are left primarily to other reference manuals (Danneman et al., 2012, Hrapkiewicz and Medina, 2007, Jacoby et al., 2002, Kohn and Clifford, 2002, Sharp and Villano, 2012). Of course, no reference text can substitute for specialized instruction and advanced training in animal physiology, handling, and medicine. Routine consultation with veterinary care professionals is strongly encouraged, particularly for any challenging or unusual clinical cases in laboratory animal species.

obtaining a clinical history and medical records

Although the individual signs can vary between species and patients, in general, health status assessments are comprised of a combination of: appearance, performance, productivity, and exhibition of appropriate and species-specific behaviors (Clark et al., 1997a, 1997b, 1997c, 1997d). Obtaining a thorough clinical "history" of the animal will be essential to integrate and analyze the relevant information necessary for critical care diagnoses and interventions, similar to what would be expected for human and pet medicine (Aldrich, 2005).

Individual medical records can be created to provide information about the condition being reported, potential research interactions with proposed treatments, action expected from the investigational team and timeframe for expected response, treatment options, and detailed instructions on treatment administration (Couto et al., 2003). Specifics about the laboratory rodent patient database ideally should include the following:

- Genetic background, strain, or stock
- Gender
- Date of birth (age)
- Body weight (BW; grams)
- Food and water provisions (type, amount)
- Caging and bedding type
- Breeding status (nulliparous, multiparous)
- Microbiological status (specific pathogen free vs. known *Helicobacter* positive, etc.)
- Enrichment provisions for exhibition of species-specific behaviors (nontoxic, disposable cotton padding, plastic tubing)
- Housing arrangement (single housed, paired, multiple cage mates) or if new cage mates have recently been introduced
- Housing parameters of the room: temperature, humidity, illumination, noise, vibration
- Whether experimentally naïve or if experimental treatments were provided
- Any adverse outcomes anticipated for the anticipated phenotype
- Recent surgical or nonsurgical procedures

- Administration of anesthesia or analgesia (type and time when delivered)
- Transportation or exposure to new housing areas
- Timeline of clinical abnormality (development, progression)
- Corrective measures (treatment plan) taken as a result of variation from normal health or behavior
- Doses, routes, and frequency of administration of any additional drugs or medications
- Resolution of the clinical abnormality (either corrected, healed, or potentially removed from study)

Assessments of the critical patient must be comprehensive, organized, focused, and efficient (Figures 1.1 and 1.2) to identify as many of the patient's problems in the order of their importance for survival (Aldrich, 2005).

Case Presentation

Date: ______________________ Facility: ______________

Species:________ Sex:________ Strain:________ Vendor: __________

Naive: ______________ Animal on Protocol: ________________

Protocol Title:

Brief Clinical History:

Problem including differential diagnoses:

Diagnostic work-up: (X-ray; CBC, Chemistries, Urinalysis, Skin Scraping, Fur Pluck, Anal tape test, Serology, Viral Culture, Culture and Sensitivity, etc.)

Progress and/or changes in treatment based on diagnostic results and/or poor progress:

Prognosis:

Necropsy/Histopathology: ______________________

Fig. 1.1 Clinical history (medical record) template for laboratory animal case management. CBC = complete blood count. (Modified from the University of Pennsylvania, ULAR.)

Sick Animal Report
Date: __/____/____ Reported by: ____________
Facility: __________ Room: ______ Cage Card # ________ Species: ____ Sex: ___
PI __________ Protocol # __________ Number of affected animals ___ /___

Name of lab contact ____________ Phone # ____________

Problem:
□ FIGHTING(□ *Separate animals*) □ SKIN LESION(not fighting)
□ PROLAPSE (rectum, penis, or vagina) □ Place on Alpha-Dri
□ TUMOR ON BACK greater than penny size or ulcerated.
□ TUMOR other than on back any size, OR multiple tumors
□ OTHER, please describe ____________

EMERGENCY* Call clinical staff immediately**
□ DEHYDRATED/HUNCHED (place 5 moist food pellets on cage floor)
□ DYSTOCIA □ TRAUMA/BLEEDING
□ MORIBUND □ PAINFUL□ BREATHING PROBLEM
CIRCLE AFFECTED AREA ON DIAGRAM TO THE RIGHT:
Mark affected area with an "X"

Ventral Dorsal

REPORT EMERGENCIES TO THE CLINICAL STAFF IMMEDIATELY
1. **WEEKDAYS:** Veterinary Technicians @ phone number XXX-XXX (M-F 7:30 am to 4:30 pm)
2. **WEEKEND:** Veterinary Technicians ALL Facilities @ phone number XXX-XXX (Sat-Sun 7:30 am to 4: 30 pm)
3. **AFTER HOURS:** Call Emergency Service to reach Vet On-Call @ phone number XXX-XXXX

Fig. 1.2 Sick animal report template for initial assessment performed cage-side. PI = primary investigator. (Modified from the University of Pennsylvania, ULAR.)

body condition scoring

Methods to develop body condition scores (BCSs) have been described in a wide variety of domesticated, herd, and laboratory animal species. The BCS serves as a semiquantitative method of assessing body fat and muscle (see relevant sections on this topic in Chapters 2 and 3). Generally, the BCS is independent of BW and frame size yet can provide a consistent approach to health assessments from appropriately trained personnel. Overall appearance and body condition score is typically ranked on a scale from 1 (emaciated) to 5 (excessive body fat), with an expectation that an animal categorized as BCS 3 is of appropriate size and species-specific conformation. Extremes in BCS may correlate with, or be predictive of, certain disease conditions. Animals with the lowest BCSs on the scale (BCS 1–2) typically are frequently monitored for any potential humane interventions necessary to improve welfare; however, it is not uncommon that the lowest BCS will result in veterinary recommendations for removal from experimental procedures due to animal welfare concerns.

relocation for physical examinations

Initial assessments of laboratory rodents will typically be performed within the original housing rooms. This assessment is usually with the housing cage removed from the rack and subsequently opened within the room's biosafety cabinet, laminar flow hood, or benchtop area. It is recommended that veterinary treatments be brought into the housing room, if at all possible, for application and administration. These types of veterinary tools can include various topical ointments (may be premixed or aliquoted into single-use vials), cotton-tip applicators, bandage materials, and sterile fluids, needles, and syringes.

If animals need more intensive care or require anesthesia for administration of the treatments, animals should be gently moved to a housing cage without cage mates or into a disposable transport container to transfer them to a dedicated procedure area or laboratory environment. Cages should be gently manipulated, with the preference for hand carriage, to avoid excessive jostling (e.g., as would happen with placement on a wheeled cart) or the possibility of dropping a cage en route to the destination.

Access to food and fluid will be necessary if the animals are away from these supplies for longer than 1–2 h; in a transport situation, a few softened feed pellets and nutritional gel supplements may be the best option for rapid administration and placement onto the holding cage floor of the ill animal.

monitoring critically ill rodents

Critical care management occurs when a clinical condition (see Table 1.1), whether spontaneous, idiopathic, or experimentally induced in laboratory rodents, may develop toward severe impairment, pain, distress, further injury, and potential for death.

Some of these situations may be more appropriately referred to as *emergencies* but will still require assessment, consultation between veterinary staff and investigators, a formulated treatment plan, and medical intervention (see Table 1.2). A representative list of examples of laboratory rodent emergencies necessitating immediate attention should include

- Blood or blood-stained discharges from any orifice
- Dehydration

Table 1.1: Diagnostic Differentials Based on Clinical Conditions Observed in Laboratory Rodents

Clinical Condition	Differentials for Consideration
Abdominal distention	Fat, ascites, splenomegaly, hepatomegaly, neoplastic mass, urinary tract obstruction, enlarged bladder, genital gland abscessation ♀ only: uterine enlargement/pyometra/mucometra, pregnancy
Abdominal pain	Gastroenteritis, urinary tract obstruction, peritonitis, abscess
Diarrhea	Gastroenteritis, parasites, gastrointestinal neoplasia, nonsteroidal anti-inflammatory drug (NSAID) administration
Nasal/ocular discharge	Conjunctivitis, corneal ulcer, glaucoma, respiratory tract disease, distress
Pallor	Shock, anemia (blood loss, potential gastrointestinal loss, leukemia, other neoplasia, marrow suppression, coagulopathy, chronic disease)
Ptyalism (salivation)	Malocclusion, oral wound, drug reaction, foreign body obstruction
Rectal prolapse	Gastroenteritis, parasites, *Helicobacter, Citrobacter,* secondary to parturition or tenesmus
Seizures	Hypoglycemia, toxicity, trauma, central nervous system mass, neoplasia, subdural bleeding, metabolic disturbance, experimental induction, expected phenotype
Stranguria	Urolithiasis, cystitis, fight wound contracture and obstruction, neoplasia, sex gland abscessation, bladder atony, paresis/paralysis
Weakness (posterior)	Systemic disorders (hypoglycemia, anemia, hypoxia), myelopathy and muscle wasting (trauma, disk disease, neoplasia, infectious)
Weight loss	Gastroenteritis, neoplasia, cardiac disease, dental disease

Source: Modified from Ivey, EI, and Morrisey, JK. 1999. *Vet Clin North Am Exot Pract* 2:471–494.

- Dystocia
- Hypothermia (cold to the touch)
- Limb weakness or paralysis
- Moribund state
- Postsurgical complications, like dehiscence of incisions
- Prolapses of eyes/urogenital organs/tissues
- Rapid weight loss, emaciation and dehydration over 24–48 h
- Respiratory distress
- Seizures that are unrelenting

Table 1.2: Rodent Health Abnormalities, with Emergencies Highlighted for Situations that Necessitate Immediate Clinical Intervention and Consultation with Veterinary Staff

Behavior/Activity	**Tumors**	**Respiration**
Hyperactivity	Tumor greater than 2 cm on back	Gasping[a]
Lethargic		Rapid breathing[a]
Head tilt	Tumor/growth anywhere else on body	Labored breathing[a]
Circling movements		
Tremors/twitching	Multiple tumors	**Oral/Nasal**
Uncoordination	Ulcerated tumor[a]	Staining
Seizures		Salivation
Continuous convulsions[a]	**Limbs/Joints**	Nasal discharge
	Swollen limb	Incisor broken
Appearance/Condition	Lameness	Malocclusion
Infection	Self-mutilation[a]	Swollen muzzle
Discolored legs, feet, ears, and tails	Missing limb[a]	Frothy discharge
	Loss use of limb/paralysis[a]	Discolored gums[a]
Distended abdomen		
Rough hair coat	**Eyes**	**Feces/Urine**
Hunched posture	Cloudy eye(s)[a]	Diarrhea
Tail missing	Moist discharge	Bloody feces
Tail swollen	Dry, crusty discharge	Feces absent
Weight loss	Red-stained eye(s)	Discolored urine
Moribund (near death)[a]	Sunken eye(s)	Urine absent
Profuse bleeding[a]	Partially closed lids/squinting	
Cold to touch[a]		**Urogenital and Anal Region**
	Eye not visible	
Skin and Hair Coat	Eyelid(s) red/swollen	Prolapsed rectum
Skin swelling	Scratching/rubbing at eye[a]	Prolapsed penis
Scabs/fight wounds	Prolapsed eye[a]	Vaginal discharge
Skin flaking		Discharge from penis
Skin discoloration	**Ears**	Prolapsed vagina/uterus[a]
Ulcerative lesion	Scabbed ear	Dystocia[a]
Trauma	Swollen ear	
Stained hair coat	Discolored ear	
Dehydration/skin tenting[a]	Discharge from ear	
	Torn ear	

Source: Courtesy of the University of Pennsylvania, ULAR.
[a] Indicates emergency.

- Self-mutilation
- Tumor burden that is ulcerated or interferes with mobility

In a critical care situation, the initial goal is to keep the rodent patient conscious with appropriate thermal, fluid, and nutritional support. The key approach is to stabilize the critical patient and *not* to determine the immediate diagnosis (Mader, 2002). It is entirely possible to inadvertently and irreparably harm a rodent patient by trying to perform too many procedures in succession, and the struggle of the patient during these events (e.g., venipuncture, cystocentesis, or radiography) may be distressful and potentially fatal (Paul-Murphy, 1996).

Clinical evaluation and assessment of critical care parameters should be performed in any sick rodent; particularly, the tenets of airway/breathing/circulation (the ABCs of critical care) can be checked by observing respiratory rate and perfusion of tissues (e.g., color of ear tips, gums, rectal mucosa) in rodents (Flegal and Kuhlman, 2004). To better relax and temporarily restrain a stabilized rodent patient for overall physical assessments, one may consider placing the animal under anesthesia (preferably isoflurane or a similar agent delivered by nose cone connected to a vaporizer); keep in mind that the influence of anesthesia will potentially bias the objective clinical data points to be collected.

Relevant questions, similar to those asked in veterinary clinical practice (Aldrich, 2005), should be reviewed on initial physical examination of the critical laboratory rodent patient, such as the following:

- Is the patient conscious?
- Is the patient trying to breathe?
  - If not, ventilate if possible.
  - Assess mucous membrane color, which should be pink/red and not purple/blue (indicating a lack of oxygen).
- If the patient is trying to breathe, is air moving in and out of the lungs?
  - Observe thoracic wall movement.
  - Auscultate breath sounds.
- Is the heart beating effectively?
  - While the heart can be palpated for activity by gently holding the thorax between thumb and forefinger, manual assessment of beats per minute is extremely difficult.

    - Auscultation can be performed using a pediatric stethoscope.
- Is there any disabling condition (trauma, neurologic injury) that may affect outcome?
- Is the patient in shock?
    - Assess
        - Decreased mental awareness.
        - Unpigmented mucous membrane color.
        - Capillary refill time (assess using rectal mucosa in mice and rats): equivalent to the time it takes for blood to return to the capillary bed after one compresses tissue with a fingertip (Aldrich, 2005).
        - Relative heart rate (as pulses may be impossible to detect in small rodents).
        - Colder extremities versus warmer central temperature differences (Lichtenberger, 2007).
        - Consider potential for disseminated intravascular coagulation, a complex systemic disorder involving the generation of intravascular fibrin and the consumption of procoagulants and platelets; thrombin ultimately potentiates the coagulation cascade and leads to small- and large-vessel thrombosis, with resultant organ ischemia and organ failure.
- Is the animal dehydrated?
    - Gently lift and pinch the skin over the back just behind the scapula and let it return to its resting position; slowness of return is correlated to various degrees of dehydration, expressed as percentages of BW as follows:
        - *marked* speed of skin return: hydrated
        - *moderate* speed of skin return: 5–7% dehydrated
        - *mild* speed of skin return: up to 12% dehydrated
- Is the animal actively bleeding?
- Is the animal in pain?
    - Consider various behaviors that can indicate pain, including vocalization when handled, attempting to bite or escape, an abnormal posture or ungroomed appearance, and lack of movement within the cage or interacting with cage mates.

pain recognition and assessments

Guidance on assessments of types of procedures in laboratory rodents that result in painful outcomes has been reviewed (Kohn et al., 2007). Those procedures described (see Table 1.3) are commonly requested in rodent IACUC (Institutional Animal Care and Use Committee) protocols and should be considered on a continuum of potential pain and distress. Be aware that certain disease models that may also induce pain, like those that may induce inflammation or neoplasia, are described throughout the remainder of the text for assessments and interventions. The list is not to be interpreted as exhaustive, and the categorizations are only guides to assist in individual cases using professional judgment and collaboration between veterinary staff and research teams. Most, if not all, of the procedures listed are to be conducted with the animal under sedation or general anesthesia.

Animals in critical condition should be assessed for indicators of pain, particularly when they are actively involved in proposed research projects. Assessing pain and subsequent management of

Table 1.3: Categorization of Painful Procedures Performed in the Course of Research with Laboratory Rodents

Minimal to Mildly Painful	**Mildly to Moderately Painful**	**Moderately to Severely Painful**
Catheter implantation	Minor laparotomy incisions	Major laparotomy/organ incision
Tail clipping	Thyroidectomy	Thoracotomy
Ear notching	Orchidectomy	Heterotopic organ transplantation
Superficial tumor implantation	Caesarean section	Vertebral procedures
Retro-orbital sinus blood collection	Embryo transfer	Burn procedures
Superficial lymphadenectomy	Hypophysectomy	Trauma models
Ocular procedures	Thymectomy	
Multiple intradermal antigen injections		
Intracerebral electrode implantation		
Vasectomy		
Vascular access port implantation		

Source: Kohn, DF, Martin, TE, Foley, PL, Morris, TH, Swindle, MM, Vogler, GA, and Wixson, SK. 2007. *J Am Assoc Lab Anim Sci* 46:97–108.

pain may be challenging; in particular, one must be aware that each animal will respond differently to painful stimuli (Klaphake, 2006). Due to the considerable variability in pain responses, it is important that pain assessment be performed by clinicians and skilled personnel with a comprehensive knowledge of the normal behavior and appearance of the species and particular animal patient of concern (Miller and Richardson, 2011). Successive observations by a single experienced observer are likely to provide the best insight into the resolution of pain and clinical improvement of the critical patient. Unfortunately, to compound the challenge of the individual conducting the clinical observations, the sheer presence of the observer, coupled with the application of certain treatments, may also have an impact on animal behavior and muddy clinical interpretations.

Acute pain, from a known cause like an injury or surgical procedure, should be treated with analgesia. Chronic pain may be associated with subtle physiological changes that may be more challenging to effectively manage in the absence of identification of an inciting cause. Pain scales (Table 1.4) have been developed to address the unique needs of many different species, based on species-specific behavior, and should be developed for individual animal models with input from the veterinarian and IACUC and in accordance with standards of the National Research Council (NRC) guidance (Carbone, 2012, NRC, 2011, Stasiak et al., 2003).

supportive care for surgical procedures

Surgical modeling in laboratory mice and rats may include intracerebral cannula implants, heart and lung transplants, coronary artery ligations, stroke induction, device placements, radiotelemeter insertion, hepatectomies, and manipulations of other major organs. Refinements in surgery methods in rodents include the use of less-invasive (nonmidline) incisional approaches (Chappell et al., 2011) and laparoscopic procedures.

Prior to initiation of surgical procedures, it is imperative to evaluate whether the patient is at an appropriate plane of adequate anesthesia and to monitor a variety of parameters to best assess the patient's physiologic status (Morrisey, 2003). These parameters can include the following:

- *Depth of anesthesia:* Assess by direct visualization, gauging the degree of muscle relaxation, reflex response, and response

Table 1.4: Template Assessment for Pain Evaluation in Rodents

Criteria	Score	Definition
Record BW and monitor food/water intake and urine/fecal output	0	BW maintained or ↑, normal food consumption and urine/fecal output. Baseline BW is scored as 0.
	1	BW change is minor (loss of < 5% from baseline BW).
	2	Animal food consumption ↓ and water consumption variable (loss of 10–15% from baseline BW). Altered urine/fecal output.
	3	Little to no food/water intake (loss of > 20% from baseline BW).
Appearance	0	Animal "normal." Hair coat smooth, lies flat, with sheen. Eyes clear, bright, open, no discharge. Posture and movements/ambulation are appropriate for a healthy animal.
	1	Lack of grooming apparent. No other marked changes.
	2	Hair coat roughened; overall hunched appearance. Eyes and nose may have discharges or porphyrin (red) staining.
	3	Hair coat very roughened. Ungroomed. Abnormal posture. Eyes pale, sunken; closure of lids.
Measurable clinical signs: evaluate baseline respiratory rate (RR); record body temperature (BT) after all cage manipulations	0	RR (regular, even) and BT are within physiologic norms. Limbs and feet warm. Mucous membranes normal (gums and anus pink); extremities normal (ears and feet pink).
	1–2	BT may be changed by ±1–2°C; RR ↑ by up to 30% (rapid, shallow breaths; more abdominal effort).
	3	BT changes greater than ±2°C; RR ↑ by 50% or RR markedly ↓ with little visible effort.
Unprovoked behavior: determine by cage-side evaluation only	0	Normal behaviors (exploring cage, grooming, feeding); BAR.
	1–2	Abnormal behavior; less BAR; inactive when activity expected (at feeding times, at night). Guarding potentially painful area (limbs, abdomen). Twitching, lameness.
	3	Unsolicited vocalizations, self-mutilation, grinding teeth, chattering, salivating, restless, or immobile. Unresponsive, quiet.
Behavior responses to external stimuli (conduct this assessment last)	0	Behavioral responses normal for expected conditions, like being restrained for physical assessment.

(*Continued*)

Table 1.4: (*Continued*) Template Assessment for Pain Evaluation in Rodents

Criteria	Score	Definition
Behavior responses to external stimuli	1	Shows minor depression or minor exaggeration of responses.
	2	Shows moderate signs of abnormal responses; may be behavior changes (more aggressive or more docile).
	3	Animal may overreact to external stimuli, may have weak responses, or be nonresponsive.

Source: Modified from Kirsch, JH, Klaus, JA, Blizzard, KK, Hurn, PD, and Murphy, SJ. 2002. *Contemp Top Lab Anim Sci* 41:9–14.

Scoring: If a subscore of 3 is recorded more than once, then all subscores of 3 are to be given an extra point (3 becomes 4). Any subscore of 3 is potentially serious, and one should always attempt to determine the underlying cause or explanation (e.g., recently dosed with opioid, recovering from anesthesia). Consult veterinary staff if any subscores of 3 are recorded.

BAR = Bright, alert, responsive.

to stimulation (toe pinch); note that the palpebral response is lost early in a surgical plane of anesthesia, so stimuli that are more painful may be required to ensure that no response is elicited, particularly prior to initiating a surgical incision or procedure.

- *Measurement of oxygen saturation of hemoglobin (or pulse oximetry):* Equipment is available commercially for rodents but may not be on site during an emergency situation. If there are apparent oxygenation or ventilation concerns, corrections will need to be instituted (i.e., intubation, repositioning the animal to maximize lung field expansion during respiration, or tracheostomy and assisted respiration).
- *Cardiovascular performance:* Specialized electrocardiographic equipment is available commercially (but may not be on site during an emergency) for rodents to measure electrical activity of the heart, including atrial depolarization, ventricular depolarization, and repolarization.
- *Blood pressure:* Typically, indirect blood pressure measurements are obtained in smaller lab animal species; the pressure cuff is placed snugly around a limb or tail over an artery, and the monitor inflates the cuff past the point that it occludes blood flow. In general, mean blood pressure values over 60 mmHg or systolic blood pressures over 100 mmHg indicate adequate tissue perfusion.

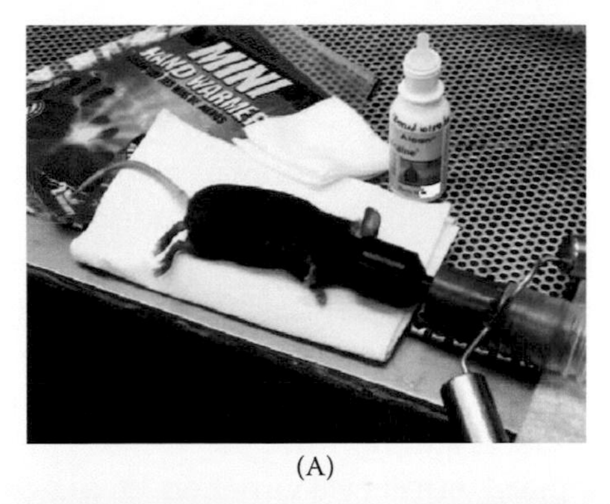

(A)

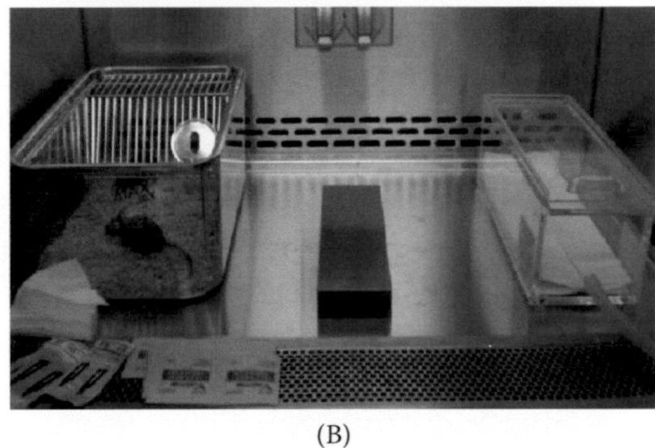

(B)

Fig. 1.3 (A) An anesthetized adult mouse in preparation for a procedure to be conducted. The funnel-shaped face mask is attached to the non-rebreathing apparatus. A downdraft table is used to capture waste anesthetic gases, and a thermogenic pack is used for supplemental heat. (Reprinted with permission from Macmillan Publishers Ltd. Yardeni, T, Eckhaus, M, Morris, HD, Huizing, M, and Hoogstraten-Miller, S. 2011. *Lab Anim (NY)* 40:155–160.) (B) The housing cage and anesthetic induction chamber can also be placed in a biosafety cabinet to conduct brief procedures that require sedation of the animal.

Pre-, peri-, and postoperative supportive care should include performance of procedures in a designated area (Figure 1.3) and provision of sterile supplies and equipment, heat through a source similar to those described (see relevant section on this topic), sterile warmed fluids, palatable calorie supplements, and analgesics. It is critical to ensure that any surgical draping allows for adequate view of animals to optimize patient monitoring. *Animals should be monitored until they are fully conscious from anesthesia and exhibiting clinically normal behaviors.*

Following surgery, it is imperative to manage potential pain and monitor anesthetic recovery, which is typically the responsibility of the research staff approved to perform the procedures. Postoperative pain in rats and mice has been shown to reduce food and water consumption; therefore, amelioration/prevention of anorexia is especially important (Flecknell, 2001). Studies have assessed the impact of socially housing rodents versus individually housing animals postoperatively; it has been shown that rats subjected to spinal cord injury have a 20% less chance of survival when housed individually, whereas socially housed female mice need less time to fully recover from telemetric implant surgery (Van Loo et al., 2007). The *Guide for the Care and Use of Laboratory Animals* (NRC, 2011) emphasizes the importance of social housing of social species; therefore, every effort should be made to socially house rodents after they have recovered from anesthesia or experimental procedures. Keep in mind that male mice may need to be housed singly; many strains are known to display behaviors suggestive of anxiety and overt aggression if exposed to unfamiliar male mice.

In the immediate hours and days postoperatively, trained personnel should frequently observe and closely monitor rodents during the recovery period, particularly to assess changes in appetite, dehydration, lethargy, or abnormal healing of surgical sites (Hoff et al., 2006). Normal circadian rhythms, as assessed in telemeterized rodents, may also take several days to return to preprocedural levels (Weinandy et al., 2005).

Once an animal has regained consciousness, the residual effects of many anesthetics may persist for up to 48 h, resulting in decreased food and water intake and prolonged ataxia (Flecknell, 1987). It is useful to record BW before and after anesthetic procedures as weight losses are typically seen during the anesthetic and recovery periods in rodents. BW may take several days to return to preanesthetic levels, even if food and water consumption appears within normal limits (Hayton et al., 1999, Lawson et al., 2001). It is often recommended to provide food and water immediately after recovery, including solid pellets on the cage floor and nutritional gel supplements, to prompt appetite following anesthesia. Typically, BW will decrease from baseline levels following anesthetic events accompanied by surgery, with a return to baseline of presurgical weights occurring within 2 weeks after the procedure (Van Loo et al., 2007). Scoring of body condition (as described previously and in forthcoming chapters) is a helpful tool to monitor the pre-, peri-, and postanesthesia appearance and health of rodents (Easterly et al., 2001, Ullman-Cullere and Foltz, 1999).

Analgesics may be necessary at scheduled intervals for several days, even up to a week or more, after the procedure. Analgesic dosing based on rodent BW tends to be relatively high due to their fast metabolism and relatively small size. Dose rates of opioids given orally are particularly elevated due to the considerable first-pass metabolism by the liver for these drugs. Dosing will vary at times, often based on strain; therefore, patients should be observed prior to and following administration of analgesics (Miller and Richardson, 2011). Further discussion of these topics is provided in other chapters.

supplemental heat provision

Rodents have relatively high metabolic rates compared to many larger domesticated species. Due to the high surface area to relatively low body mass ratio in rodents, heat loss and the development of hypothermia (colder core temperatures) are to be avoided. It is extremely important to provide thermal support at the start of anesthetic induction and continuing throughout full recovery, from anesthesia to regained consciousness (Gardner et al., 1995, Wixson et al., 1987). Body heat is lost rapidly, and this loss is accelerated when fur is removed or clipped and liquid disinfectants (particularly alcohol-based ones) are applied; in fact, animals undergoing prolonged procedures may exhibit up to 8% loss of core body heat during just 90 min of general anesthesia or prolonged sedation in the absence of appropriate supportive care. *Placement of laboratory rodents directly on surface bench tops or metal examination tables is not recommended as these can act as a heat sink and can further chill an already-hypothermic rodent.* Use of clean disposable diapers or sterile towels is preferred prior to placing the rodent patient on the examination surface.

To provide additional and supplemental warmth, warm-water recirculating blankets have been highly effective (Caro et al., 2012), and microwaveable or automated thermogenic gel packs, reflective foils, surgical drapes, and heating lamps can be used. Open-style incubators (ThermoCare®, Incline Village, NY) and Bair Hugger® blankets (Arizant Healthcare, Eden Prairie, MN) are more sophisticated heating methods that can be part of an intensive care/postoperative area in laboratory animal settings. For more impromptu methods of warming, histology slide tray warmers can be adapted in the laboratory setting; as well, the use of disposable ThermaCare® activated hand warmers (which can later be placed in the recovery cage itself) and protective latex/nitrile gloves filled with warm water can be placed near animals to provide thermal support. Care should be taken to ensure that animals do not become too warm as hyperthermia can occur easily once heat sources are provided due to the vasodilatory effects of most anesthetics (Morrisey, 2003).

For monitoring of temperatures in rodents, there are limited options available that will provide accurate assessments. Microchip transponder systems that remotely read subcutaneous temperatures have been tested (Hankenson et al., 2013) and were adequately similar in readings to traditional rectal thermometry (Chen and White, 2006, Kort et al., 1998) and intra-abdominal telemetry (Vlach et al., 2000). Noncontact infrared industrial thermometers and temporal artery thermometry have been tested on rodents but may not be consistently representative of core body temperatures achieved by other methods; human pediatric ear thermometers may be of use in larger rats, but are too large for the ear canal of most laboratory mice. The targeted temperature to which a rodent should be warmed is *95–100°F*. Keeping the body temperature as constant as possible from onset to end of anesthesia has been linked to reduction in recovery timcs by 20–30 min (Luboycski, 2008).

drug therapy

Administration of therapeutic treatments (drugs) is an integral aspect of critical care medicine. Laboratory animal patients need to be evaluated carefully prior to treatments for individual factors (whether experimentally induced or spontaneous) that can affect the distribution and metabolism of therapeutic drugs (Hackett and Lehman, 2005). It is important to be aware of the types of drugs administered, particularly with respect to the potential interactions they may have with other provided treatments. It will be essential to have a plan for the appropriate dose, frequency, route, and unique patient factors that might influence drug dosages (Hackett and Lehman, 2005). Often, for rodent patients in particular, injections are preferred to best deliver experimental and pharmaceutical agents. Understanding the different injection techniques, with regard to rate of delivery as well as volume limits, is important when choosing the appropriate injection method.

Routes of Drug and Fluid Administration

The routes of drug and fluid administration are shown in Figure 1.4.

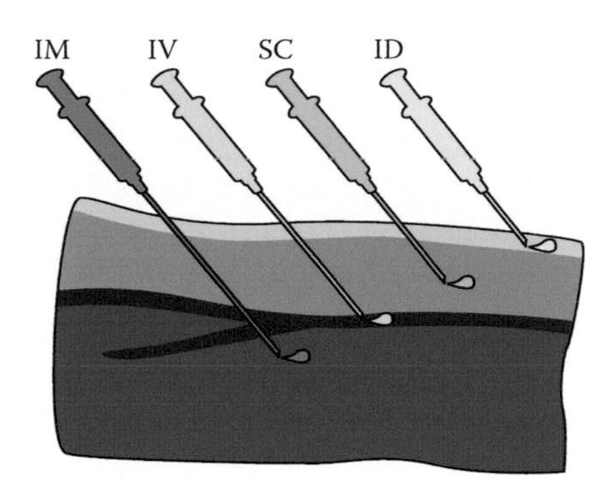

Fig. 1.4 Differing routes of administration of substances, including intramuscular (IM), intravenous (IV), subcutaneous (SC), and intradermal (ID) routes. (Reprinted with permission from AALAS. Turner, PV, Brabb, T, Pekow, C, and Vasbinder, MA. 2011. *J Am Assoc Lab Anim Sci* 50:600–613.)

- *Subcutaneous (SC).* This is one of the most common sites for drug and fluid administration in laboratory rodents. The excess skin over the back of the rodent between the shoulder blades (scapulae) accommodates boluses of injectate; volume overload is less likely with subcutaneous injections than other administration routes.
- *Intraperitoneal (IP).* Injections of fluid bolus into the peritoneal cavity are routinely performed in laboratory rodents. The needles should be directed toward the lower left or right quadrant, with a retraction of the plunger to ensure that gastrointestinal structures and the bladder have not been inadvertently penetrated. Fluids should be warmed when administered to avoid induction of hypothermia following bolus. Advantages of intraperitoneal injection include rapid absorption of large volumes of fluid.
- *Oral (PO).* Hypertonic solutions and solutions of high caloric value are often delivered by syringe placed into the diastema (space between the incisors and premolars/molars) for the animal to ingest. Oral gavage can be performed, but this may unnecessarily stress the critical patient. Small volumes are essential, with 0.05 ml often the limit per dose of what can be administered to the reluctant and critical patient (Klaphake, 2006).
- *Intravenous (IV).* Tail vessels are typically the best route in rodents for intravenous access; however, the time needed

to successfully place and secure (to prevent chewing by the patient) an intravenous catheter should be balanced against the need for immediate fluid delivery. Typically, in veterinary practice, the intravenous access is the most critical for emergencies; however, this is routinely difficult to access in clinically healthy mice and rats and more challenging in the ill rodent, which may already be volume contracted.

- *Intraosseus (IO).* Intraosseus injection of fluid permits rapid vascular access to bone marrow but should only be used in very small animals.

Note: Unless the patient is critical, very small, severely dehydrated, and hypothermic and has no accessible vessels, an intraosseus catheter should be *the last choice* for fluid administration (Mader, 2002). Intraosseus catheters are painful, are difficult to keep patent in small animals, can cause peripheral damage during placement, and can serve as a nidus for development of osteomyelitis. Common injection sites include the tibial tuberosity, femur trochanteric fossa, wing of the ileum, and the greater trochanter of the humerus.

- *Intramuscular (IM).* Intramuscular injections of bolus fluids are not recommended due to the limitations of muscle mass. For rats, injection of very small volumes into the musculature of the tongue can be rapidly absorbed due to preferential circulation to the head and brain during shock. On the contrary, certain emergency drugs, like epinephrine, may require intramuscular administration. Anterior hind leg muscles should be avoided to prevent the risk of inadvertently harming the sciatic nerve (Klaphake, 2006).

Alternative drug routes for mice and rats include intratracheal instillations (diluted doses of drugs like epinephrine, atropine, lidocaine, and dexamethasone delivered by catheter into the endotracheal tube or instilled directly via tracheostomy), gingival injections, and as a modified suppository per rectum (Hackett and Lehman, 2005). Intracardiac injections are appropriate only in cardiac arrest situations when no other method of drug administration has been successful.

Training videos are available online to review these approaches (http://www.jove.com/video/2771/manual-restraint-common-compound-administration-routes-mice; Machholz et al., 2012). Properties of the agent injected must also be considered to ensure it does not cause irritation if given via the selected technique. Certain injection sites may require restraint of animals or general anesthesia of the animal. All rodent patients under critical care should be monitored closely for adverse drug effects. Further discussion of these topics and the reference formulary for this text are provided in the final chapters and appendices.

references

Aldrich, J. 2005. Global assessment of the emergency patient. *Vet Clin North Am Small Anim Pract* 35:281–305.

Carbone, L. 2012. Pain management standards in the eighth edition of the *Guide for the Care and Use of Laboratory Animals. J Am Assoc Lab Anim Sci* 51:322–328.

Caro, A, Hankenson, FC, and Marx, JO. 2012. Comparison of thermoregulatory devices during rodent anesthesia and the effects of body temperature on physiologic parameters. *J Am Assoc Lab Anim Sci* 51:685–686.

Chappell, MG, Koeller, CA, and Hall, SI. 2011. Differences in postsurgical recovery of CF1 mice after intraperitoneal implantation of radiotelemetry devices through a midline or flank surgical approach. *J Am Assoc Lab Anim Sci* 50:227–237.

Chen, PH, and White, CE. 2006. Comparison of rectal, microchip transponder, and infrared thermometry techniques for obtaining body temperature in the laboratory rabbit (*Oryctolagus cuniculus*). *J Am Assoc Lab Anim Sci* 45:57–63.

Clark, JD, Rager, DR, and Calpin, JP. 1997a. Animal well-being. I. General considerations. *Lab Anim Sci* 47:564–570.

Clark, JD, Rager, DR, and Calpin, JP. 1997b. Animal well-being. II. Stress and distress. *Lab Anim Sci* 47:571–579.

Clark, JD, Rager, DR, and Calpin, JP. 1997c. Animal well-being. III. An overview of assessment. *Lab Anim Sci* 47:580–585.

Clark, JD, Rager, DR, and Calpin, JP. 1997d. Animal well-being. IV. Specific assessment criteria. *Lab Anim Sci* 47:586–597.

Couto, MA, Lawson, G, and Lawson, PT. 2003. Individual health reporting for mice and rats: it can be done. *Contemp Top Lab Anim Sci* 42:83.

Croy, BA, Linder, KE, and Yager, JA. 2001. Primer for nonimmunologists on immune-deficient mice and their applications in research. *Comp Med* 51:300–313.

Danneman, PJ, Suckow, MA, and Brayton, CF. 2012. *The Laboratory Mouse*, 2nd edition. CRC Press, Boca Raton, FL.

Doneley, RJ. 2005. Ten things I wish I'd learned at university. *Vet Clin North Am Exot Anim Pract* 8:393–404.

Easterly, ME, Foltz, CJ, and Paulus, MJ. 2001. Body condition scoring: comparing newly trained scorers and micro-computed tomography imaging. *Lab Anim (NY)* 30:46–49.

Flecknell, PA. 1987. Laboratory mammal anesthesia. *J Assoc Vet Anesth* 14:111–119.

Flecknell, PA. 2001. Analgesia of small mammals. *Vet Clin North Am Exot Anim Pract* 4:47–56, vi.

Flegal, MC, and Kuhlman, SM. 2004. Anesthesia monitoring equipment for laboratory animals. *Lab Anim (NY)* 33:31–36.

Gardner, DJ, Davis, JA, Weina, PJ, and Theune, B. 1995. Comparison of tribromoethanol, ketamine/acetylpromazine, Telazol/xylazine, pentobarbital, and methoxyflurane anesthesia in HSD:ICR mice. *Lab Anim Sci* 45:199–204.

Hackett, TB, and Lehman, TL. 2005. Practical considerations in emergency drug therapy. *Vet Clin North Am Small Anim Pract* 35:517–525, viii.

Hankenson, FC, Ruskoski, N, Van Saun, M, Ying, G, Oh, J, and Fraser, NW. 2013. Weight loss and reduced body temperature determine humane endpoints in a mouse model of ocular herpesvirus infection. *J Am Assoc Lab Anim Sci* 52:277–285.

Hawkins, MG, and Graham, JE. 2007. Emergency and critical care of rodents. *Vet Clin North Am Exot Anim Pract* 10:501–531.

Hayton, SM, Kriss, A, and Muller, DP. 1999. Comparison of the effects of four anaesthetic agents on somatosensory evoked potentials in the rat. *Lab Anim* 33:243–251.

Hoff, JB, Dysko, R, Kurachi, S, and Kurachi, K. 2006. Technique for performance and evaluation of parapharyngeal hypophysectomy in mice. *J Am Assoc Lab Anim Sci* 45:57–62.

Hrapkiewicz, K, and Medina, L (eds.). 2007. *Clinical Laboratory Animal Medicine*, 3rd edition. Blackwell, Ames, IA.

Ivey, EI, and Morrisey, JK. 1999. Physical examination and preventive medicine in the domestic ferret. *Vet Clin North Am Exot Pract* 2:471–494.

Jacoby, RO, Fox, JG, and Davisson, M. 2002. Biology and diseases of mice, pp. 35–120. In Fox, JG, Anderson, LC, Loew, FM, and

Quimby, FW (eds.), *Laboratory Animal Medicine*, 2nd edition. Academic Press, New York.

Kirsch, JH, Klaus, JA, Blizzard, KK, Hurn, PD, and Murphy, SJ. 2002. Pain evaluation and response to buprenorphine in rats subjected to sham middle cerebral artery occlusion. *Contemp Top Lab Anim Sci* 41:9–14.

Klaphake, E. 2006. Common rodent procedures. *Vet Clin North Am Exot Anim Pract* 9:389–413, vii–viii.

Kohn, DF, and Clifford, CB. 2002. Biology and diseases of rats, pp. 121–165. In Fox, JG, Anderson, LC, Loew, FM, and Quimby, FW (eds.), *Laboratory Animal Medicine*, 2nd edition. Academic Press, New York.

Kohn, DF, Martin, TE, Foley, PL, Morris, TH, Swindle, MM, Vogler, GA, and Wixson, SK. 2007. Public statement: guidelines for the assessment and management of pain in rodents and rabbits. *J Am Assoc Lab Anim Sci* 46:97–108.

Kort, WJ, Hekking-Weijma, JM, TenKate, MT, Sorm, V, and VanStrik, R. 1998. A microchip implant system as a method to determine body temperature of terminally ill rats and mice. *Lab Anim* 32:260–269.

Lawson, DM, Duke, JL, Zammit, TG, Collins, HL, and DiCarlo, SE. 2001. Recovery from carotid artery catheterization performed under various anesthetics in male, Sprague-Dawley rats. *Contemp Top Lab Anim Sci* 40:18–22.

Lichtenberger, M. 2007. Shock and cardiopulmonary-cerebral resuscitation in small mammals and birds. *Vet Clin North Am Exot Anim Pract* 10:275–291.

Luboyeski, S. 2008. Ensuring normothermia during general anesthesia. *J Am Assoc Lab Anim Sci* 47:113–114.

Machholz, E, Mulder, G, Ruiz, C, Corning, BF, and Pritchett-Corning, KR. 2012. Manual restraint and common compound administration routes in mice and rats. *J Vis Exp* 67, e2771, doi:10.3791/2771.

Mader, DR. 2002. Emergency/ICU procedures in small mammals, pp. 608–610. In Eighth International Veterinary Emergency and Critical Care Symposium, San Antonio, TX.

Miller, AL, and Richardson, CA. 2011. Rodent analgesia. *Vet Clin North Am Exot Anim Pract* 14:81–92.

Morrisey, JK. 2003. Practical analgesia and anesthesia of exotic pets, pp. 591–596, Ninth International Veterinary and Critical Care Symposium, New Orleans, LA.

National Research Council (NRC). 2011. *Guide for the Care and Use of Laboratory Animals*, 8th edition. National Academies Press, Washington, DC.

Paul-Murphy, J. 1996. Little critters: emergency medicine for small rodents, pp. 714–718, Fifth International Veterinary Emergency and Critical Care Symposium, San Antonio, TX.

Sharp, PE, and Villano, J. 2012. *The Laboratory Rat*, 2nd edition. CRC Press, Boca Raton, FL.

Stasiak, KL, Maul, D, French, E, Hellyer, PW, and VandeWoude, S. 2003. Species-specific assessment of pain in laboratory animals. *Contemp Top Lab Anim Sci* 42:13–20.

Turner, PV, Brabb, T, Pekow, C, and Vasbinder, MA. 2011. Administration of substances to laboratory animals: routes of administration and factors to consider. *J Am Assoc Lab Anim Sci* 50:600–613.

Ullman-Cullere, MH, and Foltz, CJ. 1999. Body condition scoring: a rapid and accurate method for assessing health status in mice. *Lab Anim Sci* 49:319–323.

Van Loo, PL, Kuin, N, Sommer, R, Avsaroglu, H, Pham, T, and Baumans, V. 2007. Impact of "living apart together" on post-operative recovery of mice compared with social and individual housing. *Lab Anim* 41:441–455.

Vlach, KD, Boles, JW, and Stiles, BG. 2000. Telemetric evaluation of body temperature and physical activity as predictors of mortality in a murine model of staphylococcal enterotoxic shock. *Comp Med* 50:160–166.

Weinandy, R, Fritzsche, P, Weinert, D, Wenkel, R, and Gattermann, R. 2005. Indicators for post-surgery recovery in Mongolian gerbils (*Meriones unguiculatus*). *Lab Anim* 39:200–208.

Wixson, SK, White, WJ, Hughes, HC, Jr., Lang, CM, and Marshall, WK. 1987. The effects of pentobarbital, fentanyl-droperidol, ketamine-xylazine and ketamine-diazepam on core and surface body temperature regulation in adult male rats. *Lab Anim Sci* 37:743–749.

Yardeni, T, Eckhaus, M, Morris, HD, Huizing, M, and Hoogstraten-Miller, S. 2011. Retro-orbital injections in mice. *Lab Anim (NY)* 40:155–160.

2

critical care management for laboratory mice

introduction

The continued demand for laboratory mice as biomedical models of disease necessitates the refinement of diagnostics and treatments for this species. Further, many health conditions and unique strain-specific behaviors in mice can be monitored and managed for improved animal and overall colony health (Bothe et al., 2005). Development of mouse strains and maintenance of experimental models require significant investments of research funds and intellectual capital put toward specific medical model discovery and progress. Therefore, emphasis will be placed on means to promote longevity of individual animals in lieu of postmortem diagnostics, if at all possible. General information about working with laboratory mice is best reviewed in the companion text, *The Laboratory Mouse* (Danneman et al., 2012). Further background information on strains, stocks, and genotypes can also be obtained by visiting the originating vendor source web-sites; additional resources are highlighted in Chapter 5.

overall assessments

When assessing a laboratory mouse, it is essential to obtain as much information as possible about the animal and its use in research to gain the greatest portfolio of information prior to finalizing differential diagnoses. The complement of information (the clinical and experimental "history") should include the background strain of the mouse; the origin (vendor and pathogen status); any genetic manipulations (which may manifest as phenotypic abnormalities); gender; age; and specifics about experimental manipulations. When considering impacts on animal health in the research setting, knowledge of physical factors like phenotype, body weight (BW), body condition score (BCS), and potential illness related to the model are paramount.

Environmental factors may also be a confounding factor for research studies and a great influence on health, specifically factors in the macroenvironment (e.g., housing room) and those in the microenvironment (e.g., the housing cage). Macroenvironmental influences include light cycle and intensity, noise or vibration exposure, and fluctuations in room temperature and humidity. In the microenvironment, the type and volume of diet and water sources, frequency of cage changing and types of bedding substrates, and fluctuations in humidity, temperature, and air changes within the cage are to be considered with respect to maintenance of animal health. It has been shown, particularly in static housing cages, that the temperature and humidity can be significantly elevated for prolonged time periods following autoclaving and heat sterilization practices (Ward et al., 2009). Therefore, precaution should be taken to ensure animals are placed in cages with bedding approximating the cage ambient temperature to avoid exposures to heat and hyperthermia, which can be rapidly fatal.

general medical approaches to physical examination and health assessments

When presented with a critically ill mouse, to minimize stressors on the animal, it may be best to prioritize actions and *not* try to tackle everything clinically at once. Breaking up the diagnostic and treatment steps into smaller stages has been shown to subject animals to less stress and lead to improved survivability in exotic animal species; in other words, a tentatively diagnosed live patient is preferable to a confirmed diagnosis at necropsy (Doneley, 2005).

Acquiring a clinical history of the rodent patient is not unlike that done for any patient in clinical veterinary practice (see details in Chapter 1), and template medical record sheets and sick animal reports are provided (Figures 1.1 and 1.2) for mice noted to be in less-than-optimal health condition.

TABLE 2.1: MISCELLANEOUS PARAMETERS FOR THE LABORATORY MOUSE

Parameter	Value
Lifespan	2–3 years
Age of sexual maturity	6 weeks
Gestation	19–21 days
Adult body weight	28–40 g[a]
Blood volume	76–80 ml/kg = 2.3–2.4 ml total for a 30-g mouse
Food intake	12–18 g/100 g BW/day
Water intake	15 ml/100 g BW/day
Packed cell volume (PCV)	38.5–45.1%
Glucose	106–278 mg/dl[b]
Body temperature (rectal)	36.5–38.0°C (97.5–100.4°F)
Respiratory rate	80–230 breaths/min
Heart rate	500–700+ beats/min

Source: Adapted from Banks, RE, Sharp, JM, Doss, SD, and Vanderford, DA. 2010. Mice, pp. 73–80. In *Exotic Small Mammal Care and Husbandry*. Wiley-Blackwell, Ames, IA; Danneman, PJ, Suckow, MA, and Brayton, CF. 2012. *The Laboratory Mouse*, 2nd edition. CRC Press, Boca Raton, FL; and Suckow, MA, Danneman, P, and Brayton, C. 2001. *The Laboratory Mouse*. CRC Press, Boca Raton, FL.

[a] Weights will vary depending on diet, age, stock or strain, gender.

[b] Values are dependent on collection method and may be influenced by anesthesia.

Typical values for biologic parameters in mice are presented in Table 2.1. The size of the typical adult laboratory mouse ranges from about 28 to 40 g, with less distinction by weight between gender than other species; this small size makes it difficult to precisely quantify body temperature, heart rate, and respiration rate without the use of telemetric implants or other specialized equipment.

Physical Examination

Familiarity with the appearance of a routine clinically healthy mouse is key to ensure recognition of one that develops abnormal clinical signs. Visual examination of the animal is the most critical step in assessing the overall physical condition of the laboratory mouse. Observation of the animal in its home cage environment is critical prior to performing a physical examination; this permits overt lesions, behavioral abnormalities, and general activity to be assessed rapidly. Animals in poor health will likely benefit from placement in warming incubators or containers where warmth and oxygen (flow rate 1–2 L/min) may be administered automatically (Klaphake, 2006).

Prior to manual restraint and handling of laboratory mice, disposable nitrile/latex gloves should be donned. Gentle single-handed restraint

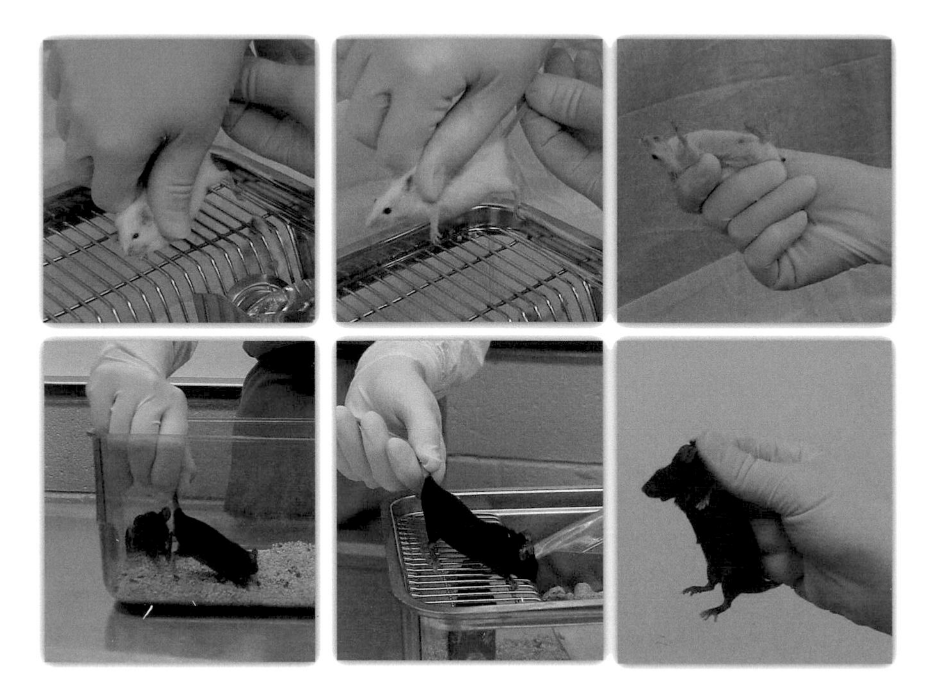

Fig. 2.1 Retrieval of mice from a cage can be conducted using a grip at the tail base to lift the animal or gently cupping the whole animal (not shown) and placing it on the cage lid. The animal can then be held by the scruff and turned on its back to better perform palpation of the abdomen for abnormalities or held upright for injections or gavage. (Images courtesy of University of Michigan, ULAM.)

of the mouse (Figure 2.1) will allow for the ability to closely observe skin and hair coat conditions, any ocular discharge or abnormalities, tooth overgrowth, abnormal masses, or unusual presentations in the anogenital region. Keep in mind that animals in critical condition may need to be sedated to perform these assessments and mitigate stress levels. Gentle palpation of the abdomen, using a pincer technique with the thumb and forefinger, should help to confirm pregnancy in females that may present with dystocia and to identify abnormalities like growths, enlarged lymph nodes (lymphadenopathy), or bladder distention. A nonpregnant abdomen is generally "soft and doughy" to the touch, and one has difficulty defining structures as particular organs; for example, something firm in the distal colon that is not consistent with fecal pellets may require further diagnostics (Klaphake, 2006).

Physiological aspects, like *body weight*, *activity*, and *behavior* assessments, are useful to measure and monitor serially. *Hair coat quality* should be reviewed regarding location areas of alopecia (baldness), open or closed wounds, or poor grooming. In addition, *respiratory status* (difficult or labored breathing with a more frequent/ diminished rate than expected) should be evaluated. Relative *perfusion status*, ascertained by the color of mucous membranes, reflects the transport of fluid and oxygen in blood to meet metabolic needs. For a mouse, perfusion can be most easily ascertained by rectal mucosal inspection; as well, the color of nonpigmented ears and tails may assist with this assessment. Collectively, these physiologic measures provide a crude interpretation of the "ABCs" (airway/breathing/circulation) of critical care medicine. Finally, the particular experimental use of the mouse, as described and approved in an Institutional Animal Care and Use Committee proposal, must be considered, and any adverse effects of the experimental procedures should be documented.

Body Condition Scoring

Assessing general body condition, as a means for assessing health status in any animal species, is an excellent tool to apply toward mice. The use of a BCS (body condition score) scale (generally on a range from 1 [wasted; emaciated] to 5 [obese]) is greatly enhanced by the adherence to definitions of numerical health diagrams that represent each score on the scale (Ullman-Cullere and Foltz, 1999). This tool is exceptionally valuable for any laboratory animal group with variable experience in working with mice, as it provides a uniform health assessment tool (Figure 2.2).

Overall percentages of weight loss should be monitored, yet may or may not be pertinent, depending on the disease model and whether the animals are expected to or may spontaneously develop tumors (Paster et al., 2009) or ascites. Typically, weight loss of more than 20–25% from preexperimental baseline may warrant critical care measures and potentially euthanasia, depending on institutional policies.

Clinical Assessments of Ill Health and Pain in Mice

Mice are prey species; as such, they are conditioned to suppress overt painful and ill behaviors, particularly when being handled by personnel. Clinical assessments of ill health and pain in mice have

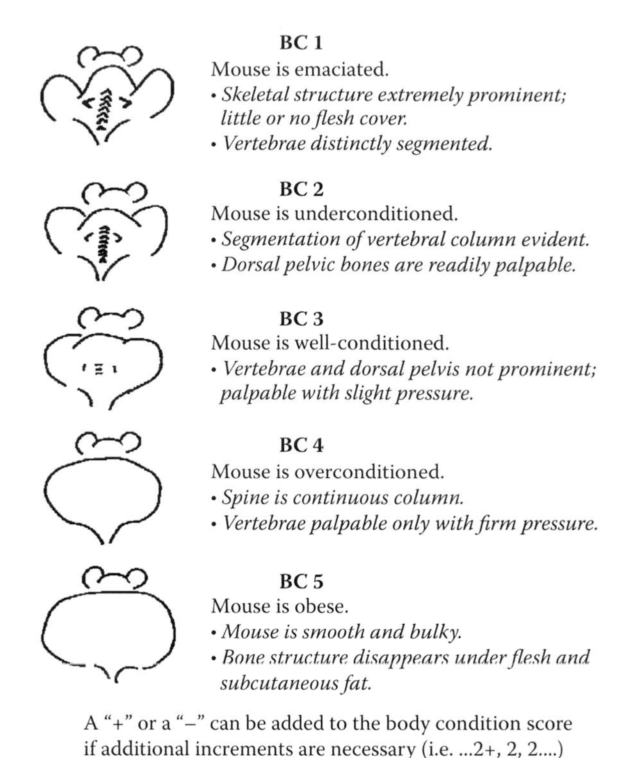

Fig. 2.2 Schematic for scoring mouse body condition. (Reprinted with permission from AALAS. Ullman-Cullere, MH, and Foltz, CJ. 1999. *Lab Anim Sci* 49:319–323.)

been described (Kohn et al., 2007, Miller and Richardson, 2011) and include the following:

- Vocalization, particularly when handled or a painful area is palpated
- Reduced grooming or piloerection, leading to a "ruffled fur" appearance
- Reduced level of spontaneous and exploratory (sniffing, rearing) activity to the point that mice may not be moving: a "moribund condition"
- Hunched posture, with potential "guarding" of abdomen and reduced mobility

- Squint-eyed appearance (either unilateral or bilateral)
- Increased aggressiveness on handling; may bite without warning
- Distanced from cage mates
- Reduced BCS, likely secondary to reduced nutritional intake or experimental model resulting in muscle wasting and weight loss
- Self-mutilation (excessive licking, biting, scratching) of the painful area
- Discharge originating from the eyes, nose, ears, or perineal region
- Abnormal postures, ataxia, circling, and raised tail position
- Palpation of unexpected masses

Monitoring Frequency

A detailed and descriptive plan for scheduled monitoring of research animals both before and after an experimental procedure, including the provision of therapeutic treatments and supportive care, should be included in the IACUC protocol submission. Investigators should be aware that, as the potential for pain/distress in animals rises, there should be an increasing intensity of monitoring and frequency of observations performed.

Objective Scoring Systems

Professional and clinical judgments are inherent to the evaluation of an animal's well-being and are essential to the ultimate decision for administration of treatments or planned removal from studies for euthanasia, based on welfare rationale. To standardize approaches for care and treatment, objective data-based approaches to predicting demise and imminent death, when developed for specific experimental models, should facilitate the implementation of timely euthanasia before the onset of clinically overt signs of moribund state (Toth and Gardiner, 2000). Individualized scoring systems are a common means by which humane interventions/endpoints can be defined and implemented. Examples of template scoring systems (Figure 2.3; Table 2.2), by which the health of laboratory rodents can be judged, are provided (Adamson et al., 2010,

Scoring	Characteristics
Active curiosity (score = 1)	• Moving quickly around cage • Frequent standing at sides of cage and close to cage mates • Active investigation of surroundings
Mildly decreased activity (score = 2)	• Reduced movement around cage • Little to no investigation of surroundings • Seeks shelter/prefers corners of cage • Will move around cage when stimulated by gentle handling
Severely decreased activity (score = 3)	• No movement around cage • May be moribund or only move slightly when stimulated by gentle handling • Typically isolated from cage mates

Fig. 2.3 Template scoring system for overall rodent activity in the home cage; a score of 3 over the course of 24 h would indicate potential removal from a study and likely euthanasia.

Hankenson et al., 2013). To eliminate the inherent subjectivity in clinical assessments, it is advised that a consensus on overall health and activity scoring be established through hands-on training of observers by the principal investigator or veterinary staff.

A novel method of potential pain assessment in laboratory mice is through coding of facial expressions of pain, also referred to as the Mouse Grimace Scale (MGS) (Langford et al., 2010). The technique was adapted from the use of facial expressions in human infants; in mice, orbital tightening, nose bulges, cheek bulges, and changes in ear and whisker position are the five "action units" scored on a 0–2 scale for their prominence in still photographs taken from digital video of mice in either a baseline or pain condition due to an irritant (Figure 2.4). The MGS displays high accuracy and reliability, can quantify pain of duration up to about 1 day, and is sensitive to detecting weak analgesic effects. Dedicated observation of cohorts of laboratory mice resulted in the identification of these subtle distinctions, like the bulging alteration from the typical smooth appearance of the nose and cheek in control animals. Use of the grimace scale has been applied to mice receiving analgesia to determine efficacy of pain relief (Matsumiya et al., 2012).

Table 2.2: Visual Assessment Scoring System that Also Can Be Used to Score Animals Postprocedurally

Score	Hair Coat	Eyes	Coordination and Posture	Overall Condition
0	Normal, well groomed, smooth, sleek	Open, alert	Normal	Normal
1	Not well groomed	Squinted	Walks awkwardly or slightly hunched, otherwise active, mobile	Roughened appearance but otherwise activity and behaviors within normal limits
2	Rough hair coat, unkempt surgical site	Squinted to closed	Hunched, abdominal stretching observed, reduced activity	Slightly depressed, poor appearance, behaviors not within normal limits due to agitation
3	Very rough hair coat, hair loss, unkempt surgical site	Closed	Hunched, stumbles when moving, inactive	Depressed, increasingly poor appearance, abnormal behavior
4			Hunched, not moving	Unresponsive

Source: Modified from Adamson, TW, Kendall, LV, Goss, S, Grayson, K, Touma, C, Palme, R, Chen, JQ, and Borowsky, AD. 2010. *J Am Assoc Lab Anim Sci* 49:610–616.

Characteristics of the animal (hair coat, eyes, coordination, and overall condition) are scored independently and then summed and averaged to obtain a final pain index score.

veterinary care measures

Administration of Fluids

Evidence of dehydration has been documented as an outcome of potential pain and distress, related to underlying clinical conditions, experimental interventions, or husbandry parameters. Dehydration may be assessed by performing a skin tent or gentle pinch of scruff over the scapulae of the mouse and assessing the time that passes for the skin to return to normal placement. A prolonged return time indicates a degree of dehydration that should be ameliorated. *Any administered fluids should be from a sterile source to avoid introduction of infectious disease agents.* In the critical rodent patient,

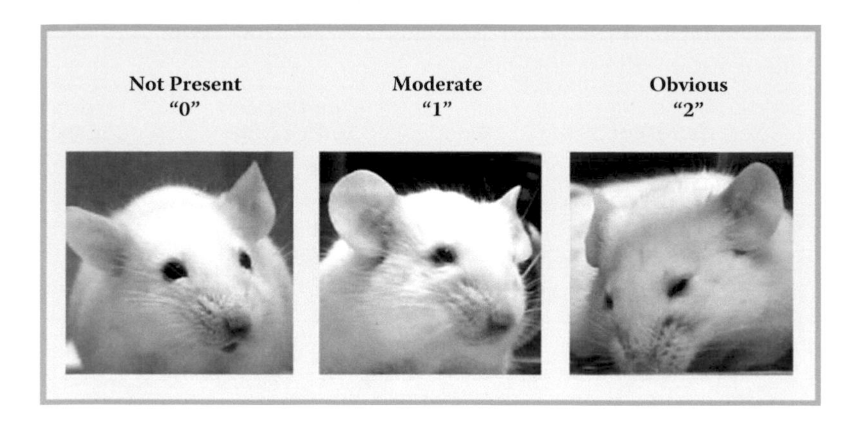

Fig. 2.4 Representative photographs of the Mouse Grimace Scale for a mouse at baseline (facial grimacing not present, 0); a mouse with moderate facial grimacing (1); and a mouse with obvious facial grimacing (2). (Reprinted with permission from AALAS. Matsumiya, LC, Sorge, RE, Sotocinal, SG, Tabaka, JM, Wieskopf, JS, Zaloum, A, King, OD, and Mogil, JS. 2012. *J Am Assoc Lab Anim Sci* 51:42–49.)

the subcutaneous (SC) route of administration may be most readily accessed; a balanced electrolyte solution (potentially with 2.5% dextrose) at an initial rate of 40–50 mg/kg twice daily is recommended and can be administered through a butterfly needle and attached line to allow less restraint of the animal (Klaphake, 2006). Prophylactic sterile fluid administration, 0.9% NaCl or polystarch given subcutaneously at a dose of 1.0 ml, can significantly improve survival rates in mouse models of cancer receiving carcinogen treatments (Smith et al., 1999). Additional fluid support is beneficial in terms of raising arterial blood pressure, although there may be a negative impact on changes in organ water content and increased potential for anemia (Zuurbier et al., 2002).

Water and fluid replacement sources are gaining in popularity, expanding from products initially developed as sustainable fluid sources for the duration of rodent shipping and transport. The provision of these water replacements, in disposable single-use containers, is typically done on the cage floor for rapid access by those animals in ill health. These supplementary fluid sources, when combined with food, can maintain the health of rodents for at least 7 days in the absence of routine water sources (Luo et al., 2003). Additional critical care considerations for nutritional support, fluid administration, and available products are provided in Chapter 4.

Blood Sampling

Blood collection, or venipuncture, is a common procedure performed in animal research for experimental, routine, or critical care reasons (Danneman et al., 2012, Hoff, 2000, Hrapkiewicz and Medina, 2007, Suckow et al., 2001). Sampling allows for testing of serum chemistry parameters, as well as complete blood counts (CBCs). Sampling sites in mice (Table 2.3) include the retro-orbital sinus (typically with animals under anesthesia), facial vein (with animals conscious), medial and lateral saphenous veins, and tail vessels (Horne et al., 2003). Retro-orbital bleeds are readily performed using microhematocrit capillary tubes formulated for collection of microvolumes of blood with appropriate anticoagulants. It has been shown that a drop of 0.05% ophthalmic proparacaine hydrochloride solution, directly onto the eye to be bled significantly reduces the incidence of responsiveness to retro-orbital blood collection (Taylor et al., 2000).

Mandibular bleeds (also referred to as using the facial vein, superficial temporal vein, submandibular, or cheek bleed) can successfully be performed with puncture of the vascular bed using a 20- to 22-gauge needle or small disposable lancet. These are recommended in mice weighing more than 20 g (Figure 2.5).

Table 2.3: Recommended Sampling Sites and Related Information for Blood Collection in Mice

Anatomical Site	Anesthesia?	Approximate Range of Volume Collected	Comments
Lateral tail vein	Not required	Up to 1% of BW	
Lateral saphenous vein	Not required	Up to 1% of BW	
Tail clip	Not required	Up to 1% of BW	~1 mm of distal end of tail should be clipped
Retro-orbital vasculature	Required	Up to 1% of BW	Should alternate eyes for repeated sampling; topical anesthetic drops prior to bleeds; ophthalmic ointment should be applied following bleed
Submandibular vessels	Not required	Up to 1% of BW	Limited to adult mice; single sampling recommended
Cardiac	Required	500+ µl	Terminal procedure only

Source: Modified from University of Pennsylvania, ULAR.

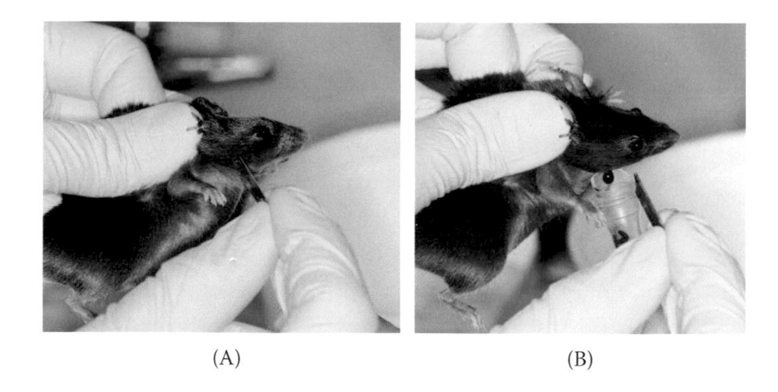

(A) (B)

Fig. 2.5 The mandibular bleeding method is used to collect samples typically from unanesthetized mice. Mice are manually restrained by a one-handed grip, and a disposable lancet or needle tip is then used to prick the facial vessels (A). Drops of blood can be collected directly into a microhematocrit tube for processing (B).

Anatomic differences in the skull between mouse strains may make it difficult to consistently find the vessels for access. Care must be taken with any method to ensure that structures (globe of the eye, ear canal, muscles, nerves, and bones) surrounding the sampling site are not injured; this may be better accomplished by inserting the flat edge of the lancet parallel to the masticatory muscle (Forbes et al., 2010). Typically, bleeds from the facial vein are recommended only for one-time, nonserial collections (Coman et al., 2010, Tomlinson et al., 2004). Nude mice have been documented as more susceptible to hematoma formation and tissue damage following sampling from the submandibular location (Nugent Britt et al., 2011). It is imperative that manual pressure be applied to achieve hemostasis following blood collection. Controversies surrounding facial sampling revolve around whether anesthesia should be used for this method. As well, facial sampling has the potential for adverse outcomes, including rapid clotting of blood, which prevents accurate serology results; formation of hematomas; blood exiting through a pierced ear canal; punctures into the oral cavity; inconsistency of volume acquisition; and high potential for pain and distress, leading to stupor and ataxia if conducted improperly. As well, there is a rare chance for significant and life-threatening hemorrhage to occur (Forbes et al., 2010).

For critical animals, it may be useful to access the saphenous vein on the hind limb or employ the tail-clip method (Figure 2.6), by which the very end of the tail is removed to allow for a drop of blood to be

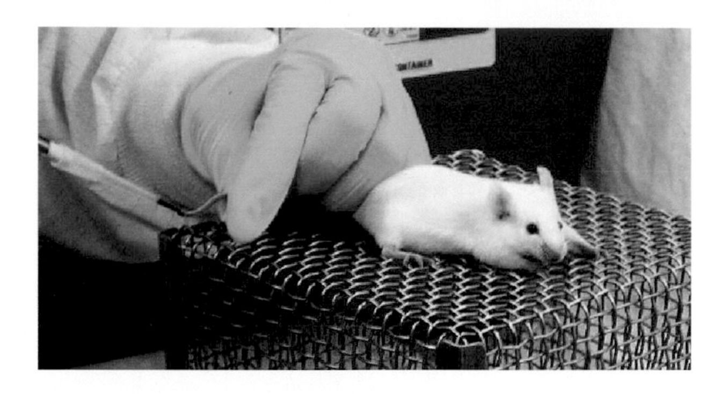

Fig. 2.6 Modified tail-clip technique. Demonstration of the minimal restraint used during the modified tail-clip blood collection procedure and the ability to pipette blood directly from the tail. (Reprinted with permission from AALAS. Abatan, OI, Welch, KB, and Nemzek, JA. 2008. *J Am Assoc Lab Anim Sci* 47:8–15.)

collected (Abatan et al., 2008). For further description, in the modified tail-clip technique, the mouse is placed on a wired surface to allow for gripping by toenails.

During the tail-clip collection procedure, the animal handler moves with the animal and only restricts animal movement if the mouse attempts to escape from the working surface. Only the distal 1 to 2 mm of the tail need to be clipped, and a capillary pipette (flushed with an anticoagulant like EDTA [ethylenediaminetetraacetic acid] or heparin) is then used to collect 20-μl samples from the exposed tail tip. Immediately after collection, styptic powder (Kwik-stop®) can be applied to the tail tip for hemostasis. For repeated sampling, the surface of the clip site can be disrupted without the need to remove additional tail tissue (Abatan et al., 2008).

Total blood volume (TBV) in a mouse has been defined as about 6–7% of BW (Hrapkiewicz and Medina, 2007, Raabe et al., 2011). Because blood regeneration rates may vary among mouse strains, a safe guideline for maximum volume of blood to remove from otherwise-healthy animals is 1% of BW (Klaphake, 2006, Paul-Murphy, 1996). For example, 0.30 ml (or 300 μl) can be withdrawn at a single sampling from a 30-g mouse; importantly, this volume may be more than needed and should be diminished to the minimum required for obtaining test results in the critical patient. Following sampling of 1% BW volume, replacement fluid therapy (0.5–1.0 ml SC or IP [intraperitoneally] of sterile isotonic fluid) should be provided.

If blood is to be collected serially (e.g., over a period of days), then sampling volumes should not exceed about 20% of BW within a 1-week period to avoid weight loss and anemia. Anemia generally is defined as a red blood cell (RBC) count, hemoglobin concentration, or hematocrit (HCT) value lower than 2 standard deviations (SDs) below the mean of a normal population (low end of normal range for HCT is ~35%); hemoglobin has been highlighted as the most direct and sensitive measure for detecting anemia at 2 SD below the mean of baseline hemoglobin values (Raabe et al., 2011). Permissible frequencies have been further delineated for 10- to 14-week-old clinically healthy C57BL/6 mice for collection of weekly blood samples for up to 6 weeks: about 15% of TBV weekly from males is acceptable, and up to about 25% of TBV weekly from females is acceptable, without adverse effects (Raabe et al., 2011).

When dealing with a sick mouse, it will be critical to decide the most informative or essential tests to run with limited volumes of blood. Blood smears can be made with fresh whole blood and the remainder of the sample collected in anticoagulant for select serum chemistry and further testing (Wiedmeyer et al., 2007).

Body Temperature Monitoring

Body temperature monitoring in mice, while a useful aspect of the diagnostic profile, is most often logistically challenging to perform. Often, simple handling of sick mice will provide some indication of whether they are excessively cool or warm to the touch. Mechanisms of obtaining temperatures in rodents can include contact methods (rectal, surface probes) and indirect methods (telemetry devices, implanted microchips, infrared laser).

Rectal probes require gentle placement and positioning during procedures in sedated mice, but are efficacious for monitoring temperature (McCann and Mitchel, 1994). Telemeterized animals, or those with an implantable device typically surgically inserted into the peritoneal cavity, are most easily monitored for variations outside the normal body temperature interval of about 95°F to 100.5°F; however, microchip transponders (Bio Medic Data Systems, Seaford, DE), surface temperature probes, and infrared noncontact thermometers may also be suitable for taking temperatures in certain mouse strains and for assessment of disease model progression (Byrum et al., 2011, Hankenson et al., 2013, Miller and Haimovich, B., 2011, Newsom et al., 2004). Placement of animals on circulating water blankets (Figure 2.7) and microwaveable padding (Figure 2.8)

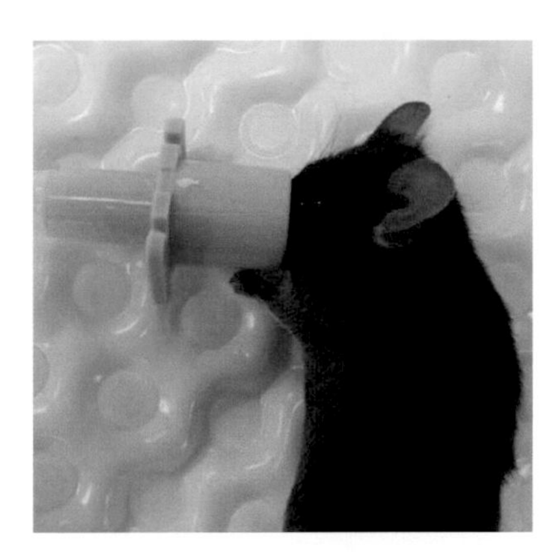

Fig. 2.7 Anesthetized mouse on warm-water recirculating blanket (T/Pump® Classic TP650, Gaymar®) placed on manufacturer's setting of "medium" heat. Animal's eyes were treated with topical ophthalmic tears to maintain moisture. (Image courtesy of the University of Pennsylvania, ULAR.)

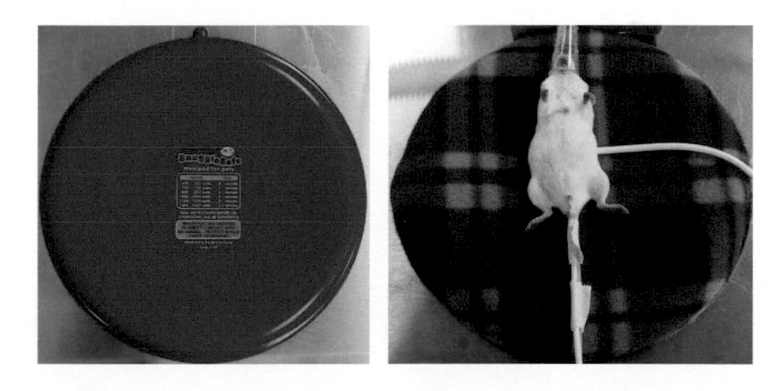

Fig. 2.8 Reusable heating pad (SnuggleSafe® Microwave Heatpad, West Sussex, UK) used for thermal support of anesthetized mice without a cover (left) and with manufacturer's cover (right). (Reprinted with permission from AALAS. Taylor, DK. 2007. *J Am Assoc Lab Anim Sci* 46:37–41.)

has been successful for maintaining body temperatures with rodents under anesthesia (Caro et al., 2012, Charles et al., 2005, Taylor, 2007).

Body temperatures may vary depending on the biomedical model. For example, in mouse models of septic shock (Nemzek et al., 2004) and infectious herpesviral disease (Hankenson et al., 2013), non-transient drops in body temperatures are a criterion for determining humane endpoints prior to spontaneous death.

Endotracheal Intubation

Mouse endotracheal (ET) intubation (Figure 2.9) has been refined to enhance the ability to mechanically ventilate and provide inhalant anesthesia to this species (Hamacher et al., 2008, Rivera et al., 2005, Spoelstra et al., 2007, Tonsfeldt et al., 2007).

Direct visualization of the arytenoid cartilages can be performed with a handheld light source combined with an otoscope. Mice anesthetized with inhalant anesthesia (e.g., isoflurane) can be placed supine on a tilt board or an incisor bar (Burns et al., 2005) or on any sanitizable surface (plexiglass) that puts the animal at an incline (Rivera et al., 2005). Mice are restrained with a rubber band placed beneath the incisors and around the support board. Transillumination is accomplished using a fiber-optic light beam or by aiming a horizontal microscope light at the midtracheal level. The patient's tongue can be held aside with a pair of forceps shielded with polyethylene (PE) tubing or held flat against the lower jaw with the bent end of a small weighing spatula. The backlit larynx is then visualized, and a 20-gauge, 1.25-inch Teflon intravenous catheter, or species-specific ET tube, is inserted into the trachea. Verification of accurate placement is noted by presence of condensation on the tube

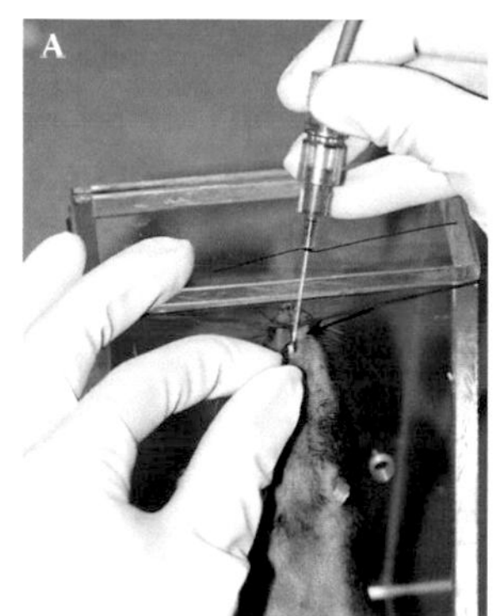

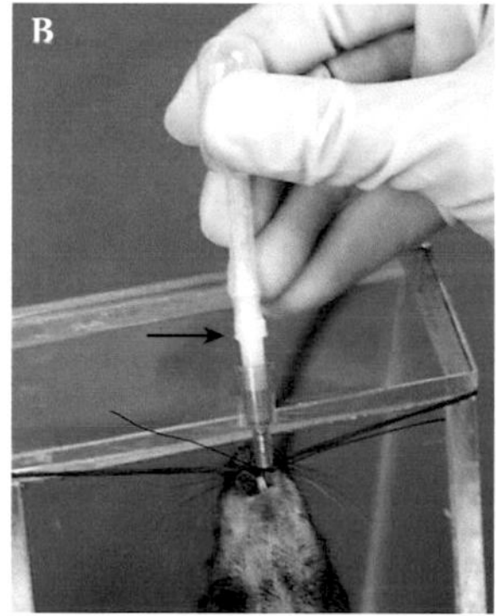

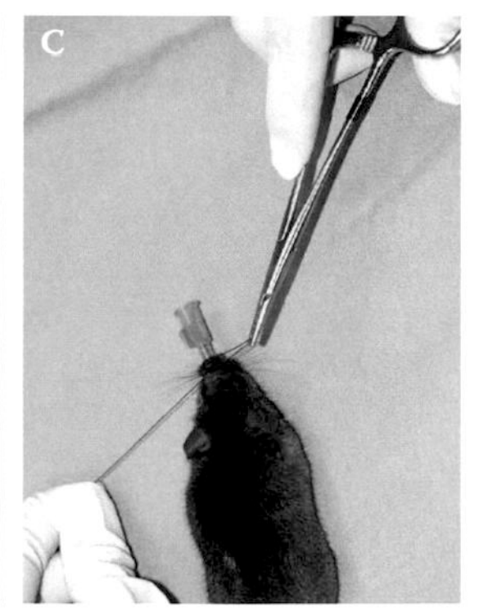

Fig. 2.9 Intubation of the mouse with (A) retraction of the tongue and insertion of the intubation system (arrow points to light at the end of the fiber), (B) verification of tube placement with a modified disposable plastic transfer pipette (arrow points to the male Luer connector), and (C) securing the endotracheal tube to the muzzle of the mouse. (Reprinted with permission from AALAS. Rivera, B, Miller, S, Brown, E, and Price, R. 2005. *Contemp Top Lab Anim Sci* 44:52–55.)

and gentle inflation of the lungs with a small disposable pipette. Once intubated, mice can rapidly be moved from the inclined support board and positioned so that the ET tube is attached to the inhalant anesthesia flow for the duration of the procedure of interest. Note that in the critically ill mouse, intubation may be extremely challenging and should be attempted only as a last resort to gain airway access if tracheostomy cannot be performed (see relevant section in Chapter 4).

Injections and Oral Administration

Injections can be performed routinely using multiple routes (Table 2.4) for the mouse, including subcutaneous, intradermal (ID), intraperitoneal, intratracheal (IT), and intravenous (IV) routes, as described in Chapter 1. As well, retro-orbital injections in mice have been described as an alternative to tail-vein injections, to deliver volumes up to 150 μl in adult animals and up to 10 μl in neonates (Yardeni et al., 2011).

Administration of drugs and fluids can also be done by oral gavage and voluntary ingestion. Creative approaches to disguising medications in palatable substances have been successful, like provision of analgesics in Nutella® chocolate spread (Goldkuhl et al., 2008, Jacobsen et al., 2011, Kalliokoski et al., 2011, Strom et al., 2012), although mice may need to acclimate to the novel substance prior to drug incorporation. Honey (in 100-μl doses administered by syringe) has been used to deliver antiparasitic medications

Table 2.4: Recommendations for Injection Dose Limits Based on Weight of Laboratory Mice

	Injection Limits (ml/kg)					
Route	**PO**	**SC**	**IP**	**IM[a]**	**IV (Bolus[b])**	**IV (Slow)**
Dose (ml/kg)	10	10	20	0.05	5	25
Weight (kg)						
0.020	0.20 mL	0.20 mL	0.40 mL	0.001 mL	0.10 mL	0.50 mL
0.025	0.25	0.25	0.50	0.001	0.13	0.63
0.030	0.30	0.30	0.60	0.001	0.15	0.75
0.035	0.35	0.35	0.70	0.002	0.18	0.88
0.040	0.40	0.40	0.80	0.002	0.20	1.00
0.045	0.45	0.45	0.90	0.002	0.23	1.13
0.050	0.50	0.50	1.00	0.002	0.25	1.25

Source: Modified from University of Pennsylvania, ULAR.

[a] Technique is discouraged in mice due to small muscle mass.

[b] A bolus is a larger dose given over a shorter period of time.

daily for up to several weeks (Kuster et al., 2012). Therapeutics delivered in peanut butter pills have been described; in brief, peanut butter is heated to 37°C, and drug doses are then mixed in for approximately 15 min. Pilling molds are used to isolate single-dose applications, which can be frozen at -80°C until needed. Mice readily consume the pellets, ensuring delivery of the complete dose (Cope et al., 2005).

Urine Sampling

Urinalyses in mice are challenging due to volume limitations; however, the propensity for mice to urinate on handling assists in the collection of free-catch samples (Chew and Chua, 2003). Urine droplets can be collected into plastic well plates and then aliquoted by pipette for appropriate assessment and assays (Table 2.5) (Kurien et al., 2004, Kurien and Scofield, 1999). Facilitating urination in the mouse can be done by applying equal pressure in a gentle massaging manner at both sides of the lower back near the tail, with the thumb on one side and the fore and middle fingers on the other side, rubbing up and down; this application of pressure to the caudal back area of the mouse facilitates expression of a maximum volume (>50 μl) of urine for collection (Chew and Chua, 2003). Alternatively, gentle, firm pressure on the bladder with the thumb and index finger can stimulate urination, which can be collected on clean plastic wrap, pipetted into sterile microfuge tubes, and stored immediately at –20°C until use (Maier et al., 2007).

Critically ill mice should be stabilized prior to attempting urinary catheterization if urine collection by other methods has been unsuccessful. Urinary catheterization should only be performed on

Table 2.5: Parameters for Urine in the Laboratory Mouse

Parameter	Value
Color	Clear or slightly yellow
Volume	0.5–2.4 mg/24 h
Specific gravity	1.030
pH	5.0
Glucose	0.5–3.0 mg/24 h
Protein	0.6–2.6 mg/24 h

Source: Adapted from Danneman, PJ, Suckow, MA, and Brayton, CF. 2012. *The Laboratory Mouse*, 2nd edition. CRC Press, Boca Raton, FL.

anesthetized mice. Aseptic technique (see Chapter 4, "Perioperative Care Considerations") and atraumatic approach should be used during placement of a urinary catheter. Prior to insertion of the catheter, the external urinary orifice should be gently cleansed using a disinfecting (e.g., chlorhexidine) solution. The individual performing the catheterization is advised to don sterile surgical gloves, use a sterile catheter, and apply a small amount of sterile water-soluble lubricant on the external urinary orifice. Additional sterile lubricant should be applied in a thin layer to cover the surface of the urinary catheter for ease of insertion into the urinary orifice (St. Claire et al., 1999). The diameter of the urinary catheter should be the minimum that can be inserted into the bladder and still prevent urinary leakage around the catheter. The distance from the external urinary orifice to the neck of the bladder should be estimated prior to catheter insertion.

The anatomy of the female mouse is unique in that the urinary orifice is external and just anterior to the vaginal opening. Catheters for adult female mice can be made by using number 10 (1.8-French) Intramedic® PE tubing. A guidewire can be threaded through the PE tubing to increase the rigidity of the catheter. Care should be taken that the tip of the guidewire does not extend past the end of the catheter. Guidewires can be made of stainless steel surgical wire (Ethicon, Somerville, NJ) and are coated with a water-soluble lubricant to ease placement and removal from the PE tubing. The approximate distance from the external urinary orifice to the neck of the bladder for a 20-g female mouse is 10 mm (St. Claire et al., 1999).

Once urine is collected, standard veterinary refractometers require approximately 60 μl of mouse urine to generate a reliable reading of the urine specific gravity value (Forbes-McBean and Brayton, 2012). Elevated plasma creatinine levels have routinely been used as a marker of reduced kidney function in animal studies; however, historically these have been difficult to measure for mice. Because blood urea nitrogen (BUN) concentration increases as kidney function declines, plasma BUN is a decent alternative to creatinine as a high-throughput screen for evaluating kidney function in mice. In addition to BUN as a marker for kidney function, the ratio of urinary albumin concentration to creatinine concentration is commonly used as an indicator of kidney damage in animal studies. It has been demonstrated that chromagens in mouse plasma do not interfere with autoanalyzer methodologies for quantifying BUN concentrations (Grindle et al., 2006).

abnormal, critical, and emergent conditions

Categories of laboratory rodent health concerns are discussed in alphabetical order to facilitate locating topical information. Under each topic, the "cause and impact" has been provided, and "potential treatments" offer suggestions about procedures, therapeutic treatments, or husbandry and environmental alterations. Every attempt has been made to provide citations from the literature for evidence-based medical outcomes. For those health concerns that list drug therapy options, please refer to the rodent formulary provided in Appendix C for additional details on dosages and route of delivery.

Abdominal Swelling

- **Cause and impact:** Animals may present acutely with an enlarged or swollen abdominal (peritoneal) cavity. This may be caused by ascites fluid accumulation (see further section if presumed research related), organomegaly, pregnancy (in females), hemoabdomen, enlarged bladder, subcutaneous edema, or neoplasia, among other differentials (Figure 2.10). The impact on the mouse can be severe due to pressure placed on the thoracic cavity, secondary respiratory difficulty due to restricted ability of the lungs to expand, and anemia due to blood loss into the abdomen.

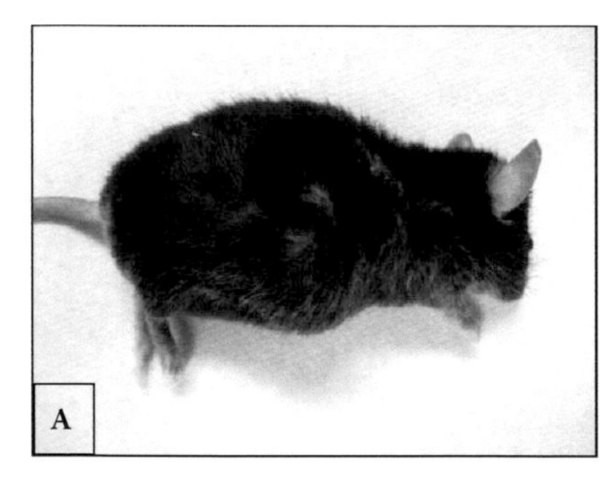
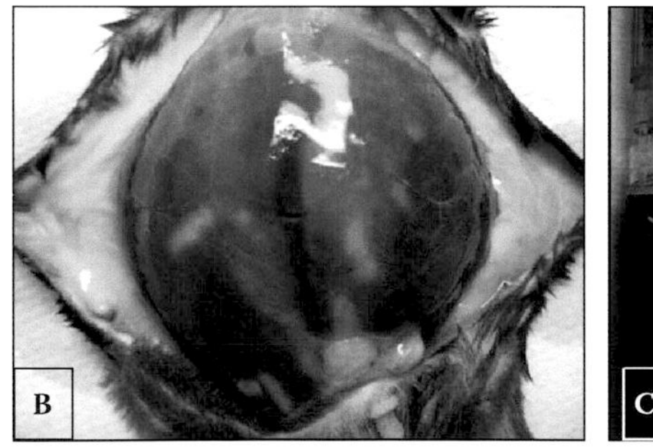
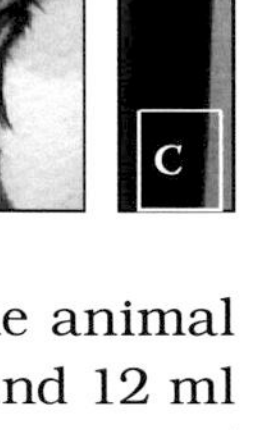

Fig. 2.10 Abdominal swelling reported in a mouse (A). The animal was subsequently found to have subcutaneous edema (B) and 12 ml of serosanguinous fluid free in the abdomen (C). (Images courtesy of the University of Pennsylvania, ULAR.)

- **Potential treatments:** Determining the differentials will be largely based on physical examination and palpation to identify masses within the abdominal cavity; however, it may be that abdominal contents cannot be appreciated due to the volume of dilation. Fine-needle aspirates conducted over the caudal ventral midline can determine if the extracted fluid is urine. Aspiration off the midline to the right, using a ventral approach, may assist with removal of free abdominal fluid, which can be evaluated for cellular and proteinaceous components (transudate, modified transudate, or exudate) by microscopic review of a fluid drop placed on a glass slide.

  Experimental information and assessment of gender of cage mates will assist with ruling out the potential for a pregnant mouse; as well, nipples may be prominent in pregnant mice.

  If the abdomen appears swollen due to a large neoplastic growth, and depending on the importance of the mouse to the research outcomes and colony, exploratory surgery may be performed to extract any growths from the peritoneal cavity. These mice will require close monitoring postoperatively, particularly due to the potential for heightened preoperative stress and weakened physiologic condition.

Abscessation

- **Cause and impact:** Coagulase-positive *Staphylococcus aureus* is commonly found on the skin of animals and has been reported as a primary cause of facial abscesses in mice (Figures 2.11 and 2.12) (Lawson, 2010). An oral route of infection has been suggested, and it has been verified that introduction of bacteria occurs through piercing of the oral mucosa by pelage or vibrissae (hairs) following grooming or barbering activities. Hair then can become entrapped in the periodontal spaces as a side effect of these activities. Severe localized periodontal bone loss in the oral cavity, secondary to hair ingestion and abscessation, has been confirmed by micro-computed tomography (micro-CT).

  Ultimately, abscess formation can occur anywhere on the body at a site where the skin integrity has been altered and bacterial contamination introduced (Figure 2.13).
- **Potential treatments:** Abscesses in mice, depending on location and size, can be treated similarly to those in other species.

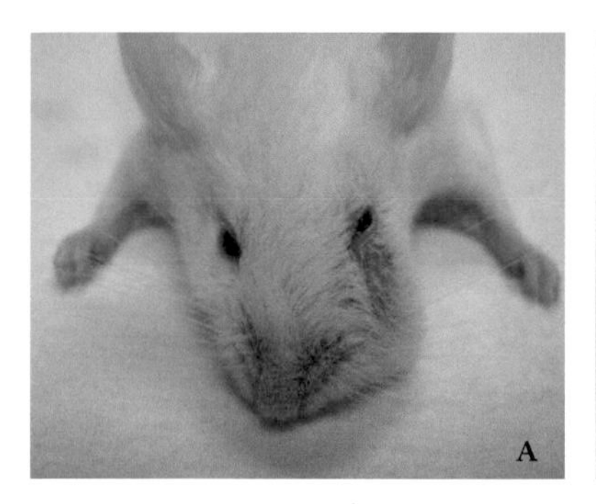

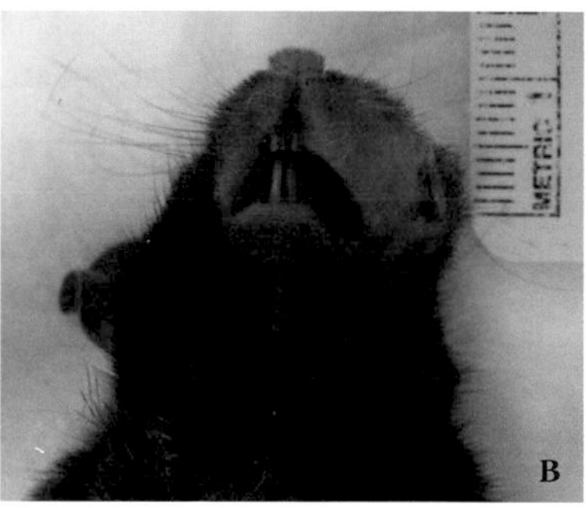

Fig. 2.11 Mouse that presented with a facial mass (A), determined to be an abscess. (Image courtesy of the University of Pennsylvania, ULAR.) (B) Facial abscess (ventral view) with associated draining tract. (Reprinted with permission from AALAS. Lawson, GW. 2010. *Comp Med* 60:200–204.)

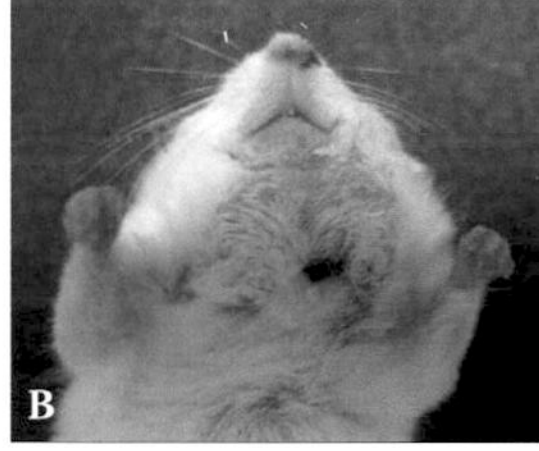

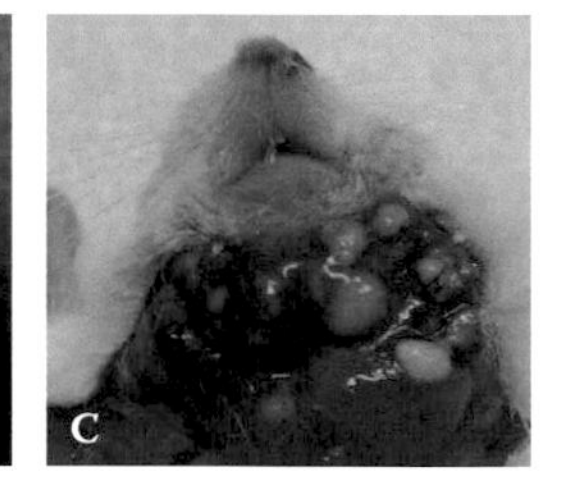

Fig. 2.12 Adult female Swiss Webster mouse housed in a conventional facility was reported with a submandibular swelling (A). Physical examination concluded that the animal was in thin body condition, hunched, with a draining tract noted on left ventral aspect of mandible (B). Degree of abscessation precluded ability to flush and drain the area (C), and the animal was humanely euthanized. Culture results confirmed infection with *S. aureus.* (Images courtesy of University of Pennsylvania, ULAR.)

Fine-needle aspiration can be performed to determine if the extracted material is purulent; culture and antibiotic sensitivity testing can be performed to best determine antibiotic treatment. Under anesthesia and using aseptic conditions, abscesses can be lanced, drained, flushed, and most often left to heal by second intention. Topical antibiotic ointment (Neosporin®; polymyxin B sulfate + neomycin sulfate + bacitracin zinc) is most commonly utilized at the site daily or every other day (EOD) for 3 to 5 days; systemic antibiotic

Fig. 2.13 Abscess draining tract in an adult female FRG (with humanized mouse liver) mouse that presented with difficulty walking. This strain is extremely immunodeficient, and the entire colony was maintained on antibiotic (enrofloxacin-treated water) to ameliorate further secondary bacterial infections. All nutritional, fluid, and caging materials were autoclaved into the room to further diminish the possibility of secondary infectious disease. (Image courtesy of University of Pennsylvania, ULAR.)

administration may be selected for additional coverage during the healing process. Analgesics should be applied to manage pain secondary to the eradication of the abscess, dependent on the level of tissue manipulation.

Cage Flooding with Subsequent Hypothermia

- **Cause and impact:** Water bottle or automatic watering system malfunctions that result in cage flooding can be a source of significant morbidity and mortality in mice used in biomedical research. Following an investigation into the cause for leaking water valves, the findings identified a combination of bedding materials and rodent fur lodged in failed water valves (Ogeka, 2009). Even relatively minor floods from watering sources into the cage can lead to significant hypothermia and potentially death if left uncorrected.
- **Potential treatments:** Prompt warming is an important first aid procedure for wet mice, with the target temperature for rewarming equivalent to 90–100°F. Since these mice are typically conscious, despite hypothermia, external warmers are advised. These can include heat lamps (250 W), microwaveable gel packs (wrapped and placed within the cage

in a location that prevents direct contact with animals), recirculating warm-water blankets, and reusable chemically activated heating pads (placed under the cage). Rewarming first aid stations can be assembled and remain permanently in housing areas to facilitate prompt treatment of any animals affected by cage flooding (Figure 2.14).

Heat lamps can be beneficial for rewarming purposes but can lead to overly hot temperatures (120°F) in less than 15 min when placed at a distance of 4 inches from cage lids. Therefore, placement of hypothermic mice, in a dry cage with bedding, at a distance of 6 inches from a heat lamp for 35 min will achieve a warming temperature of 100°F, which will then remain elevated for 60 min (Hedrick et al., 2009).

A standard rewarming practice of the University of Colorado-Denver is to scruff a damp mouse and manually submerge the animal in warm water. Water should be warm to the touch but not excessively hot. The animal is kept in the warm bath for 1-2 minutes. These mice are then dried throughly and placed in a dry and clean cage set-up.

Microwaveable pads heated to 90°F can be placed within the dry cage with mice for 10–20 min and will keep the cage

Fig. 2.14 An example of how a flooded mouse cage appears following a malfunction of the watering system (A); this rewarming first aid station (B) was developed for damp or hypothermic rodents recovering from a cage flood event, crafted from a commercial histology slide warmer (front) with a customized polycarbonate extender (rear) to support the full length of rodent cage and allow for warm and cool thermal zones. Note that the station can also be used for postoperative recovery purposes. (Images courtesy of Emory University; M. J. Huerkamp.)

warmed for 1 h; however, there is additional preparatory time associated with microwaving for this device (including having ready access to the microwave and ensuring the pad has not been overheated or contains "hot spots" from uneven heating). Reusable chemical pads can be activated manually, and they heat within 30 s. These chemical pads can be placed within a new and dry cage of mice for 20 min and will keep the cage warmed to 90–97°F for greater than 1 h (Shomer and Berenblit, 2008).

Cages exposed to heat sources should never be left unattended for prolonged periods as internal cage temperatures can reach limits detrimental to the animals (Hedrick et al., 2009).

Cannibalization

- **Cause and impact:** In rodent breeding colonies, there are often legitimate concerns about the potential for mutilation and cannibalization of neonates, resulting in the loss of valuable research animals. The causes for cannibalism are thought to be linked to stressors on the mother, including handling or disrupting neonates too soon after delivery or environmental influences, like construction noise and vibrations. Cannibalism may also be strain related in more aggressive mouse strains or may be conducted by male mice (which have fathered the offspring).
- **Potential treatments:** Husbandry practices could involve a *decreased* change cycle of cages of breeding mice with new litters, such that after parturition these cages should be left undisturbed (i.e., not changed to clean bedding) for at least 2 days postpartum. Breeding rooms can have altered light–dark cycles for maximizing production, caging materials may be tinted or colored, and enrichment materials (e.g., paper enrichment shacks and plastic tubing) can be placed in the environment to provide a degree of shielding for the dam. Providing nesting material is essential, and females will deliver pups into nests, which typically provide a softer and warmer surface (than corncob bedding) for altricial neonates during nursing and development.

Modification of poor maternal behaviors, particularly in strains known to readily cannibalize, can be attempted. Maternal administration of perphenazine on the day before or morning of parturition to dams has been reported to decrease the incidence of cannibalism in colonies of interferon-γ and interleukin (IL) 4, IL-10, and IL-12 knockout mice of the DBA/1 and C57BL/6 background strains. Perphenazine (2–4 mg/kg) can be supplied in water bottles. Results have shown that medicated dams weaned 76.4% of their pups, compared with untreated dams that weaned only 59.4% of their pups. Timing of the administration of perphenazine did not appear to have a significant impact on efficacy (Carter et al., 2002).

Nutritional supplementation with specialty treats to "distract" the dam from engaging in cannibalism have been used with success and are described further in Chapter 4, "Nutritional Therapy Considerations."

Conjunctivitis

- **Cause and impacts:** The appearance of reddened, crusty, and swollen conjunctiva in mice may be due to a number of causes and should be treated as a painful condition that should be monitored for improvement (Figure 2.15). Historically, in nude mice (Bazille et al., 2001), conjunctivitis has been linked to the contamination of the conjunctiva with cotton fibers from nesting material that results in chronic irritation. In hairless mice (SKH1) that have been reported with bilateral conjunctivitis and blepharitis, the disorder is related to body and facial hair shedding during the first weeks of the neonates' lives (Rosenbaum, 2010).

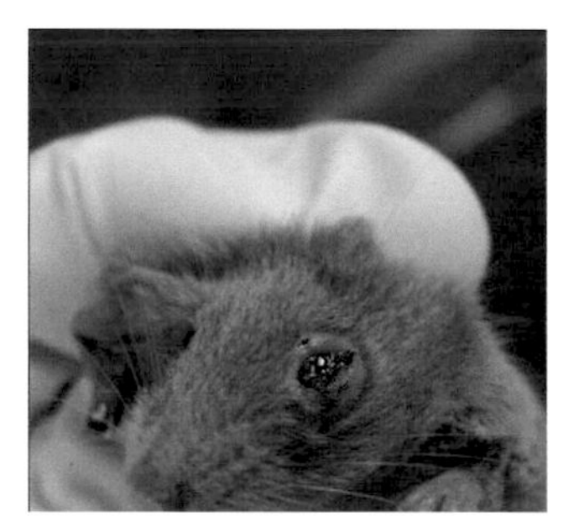

Fig. 2.15 Representative mice with presentation of severe conjunctivitis. Spontaneous development of clinical signs (left); experimental development secondary to ocular exposure to herpesviral infection (right).

- **Potential treatments:** Husbandry practices should be altered so that cages are changed with *increased* frequency during the shedding of the neonates' hairs. Known affected strains may be moved from static housing to ventilated caging as the incidence of conjunctivitis is increased in static housing conditions, likely due to the minimal airflow preventing removal of the shed hairs from the facial and ocular area (Rosenbaum, 2010).

  Routine prophylactic cleaning of the conjunctiva with sterile swabs and saline to remove fibers and hair is also effective. Removal of material embedded in the eyelids can be done with a swab primed with topical triple antibiotic (e.g., Neosporin) or a related antibiotic ophthalmic ointment (Swan et al., 2010).

Cross Fostering of Neonates/Mouse Pups

- **Cause and impact:** In the course of breeding mice for research studies, instances may arise when there is a loss of a nursing mother, either at the time of delivery or prior to weaning of the neonates or pups. The litter will not survive unless there is immediate intervention by provision of a foster mother or supplements provided by personnel.

  For experimental and biosecurity reasons, intentional cross fostering of pups can be used to eliminate potentially harmful murine pathogens, including murine norovirus (MNV), mouse hepatitis virus (MHV), and *Helicobacter* from contaminated lines of mice.

- **Potential treatments:** Placement of orphaned pups, regardless of whether they are age matched or not age matched to pups of the foster mother, has been successful for survival, with pups up to 12 days of age. Pups to be fostered should be gently rubbed with bedding (for transference of odors from the foster mother) or intermixed with the litter of pups belonging to the recipient dam to best facilitate orphaned pup acceptance (Hickman and Swan, 2011).

  To eradicate murine pathogens in newborn pups, litters should be less than 24 h old and from cages in which bedding material was changed within 24 h of planned cross fostering.

Note: *Syphacia obvelata* has not been eliminated successfully through this cross-fostering technique (Artwohl et al., 2008).

Cross-fostered mice should be tested at a minimum of 4 to 12 weeks of age to ensure specific pathogen-free status (Artwohl et al., 2008). In particular, for elimination of MNV, it has been shown that cross fostering of neonatal mice from MNV-infected to naïve dams is successful when pups are 1 to 3 days of age (Buxbaum et al., 2011, Compton, 2008). For elimination of *Helicobacter* spp., mice should be fostered within 24 h of birth (Singletary et al., 2003, Truett et al., 2000).

Milk substitutes (replacers) for artificial rearing of pups have been utilized for many years and have been based on rat milk substitutes (RMSs; derived from cow's milk) and canine pup milk replacers (Auestad et al., 1989, Hoshiba, 2004). More recently, milk substitutes have been formulated following analysis of the components of actual milk collected from ICR, BALB/c, and FVB/N mouse strains (Yajima et al., 2006). Mouse formula includes protein (purified bovine casein and whey), edible oils (up to five types, based on RMS), and vitamins similar to the RMS.

Administration of the milk replacer has been performed using gastrostomy catheters (placed surgically; Figure 2.16) and hand feeding with a surrogate nipple and nursing bottle (Beierle et al., 2004, Hoshiba, 2004).

Dystocia

- **Cause and impact:** Dystocia results from the inability of the uterus to respond to fetal signals appropriately and leads to a delay in onset or completion of pup delivery (Narver, 2012); it is one of the most common problems in rodent breeding colonies. It should not be assumed that laboratory mice delivering during daylight hours are in dystocia; delivery of pups is genetically based and may occur outside the night cycle (Murray et al., 2010, Narver, 2012). Once identified, dystocia is an emergency requiring intervention to preserve the life of pups as well as the dam. Dams with dystocia may be noted to have bloody vaginal discharge or may have pups actively lodged in the vaginal opening.

  Cross fostering (see relevant section on this topic) of surviving pups to another nursing female mouse is often required for valuable pups of mothers that decompensate during parturition and require euthanasia.

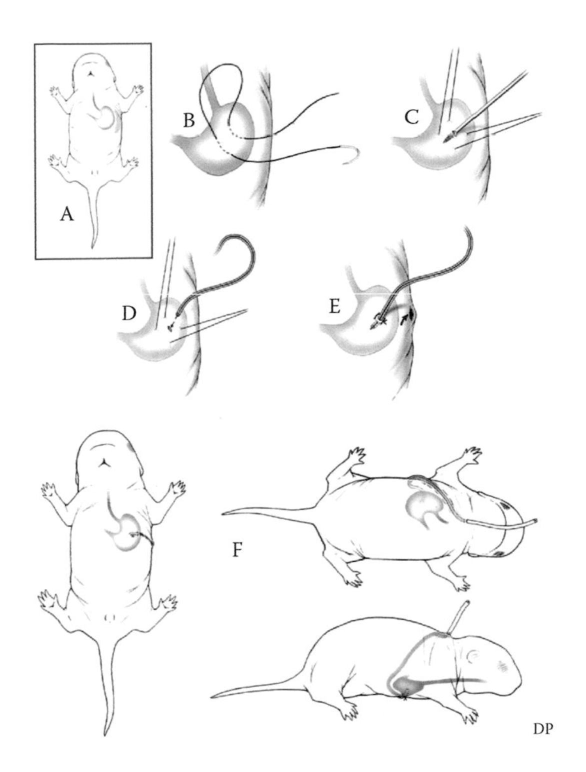

Fig. 2.16 The percutaneous insertion of gastrostomy tubes into mouse pups. (A) The mouse pup is in the ventral view, with the stomach and spleen visible. The milk-filled stomach of the pup is easily seen through the transparent skin of the anterior abdominal wall. (B) After adequate anesthesia, a small stitch in a U manner is placed through the abdominal wall, incorporating the anterior wall of the stomach, thus pulling the stomach tight to the abdominal wall. (C) A needle is used to create a small stab wound through the skin and into the anterior gastric wall. (D) A thin wire is passed through the stab incision into the stomach, and polyethylene tubing is passed over the wire into the stomach. (E) The purse-string suture is tied, and the wire and tubing are tunneled under the skin of the abdomen toward the thorax. (F) The wire and tube are tunneled under the skin of the thorax, out the nape of the neck, and the wire is removed. The final tunneled gastrostomy tube is depicted in the ventral, dorsal, and lateral views. (Reprinted by permission from Macmillan Publishers Limited. Beierle, EA, Chen, MK, Hartwich, JE, Iyengar, M, Dai, W, Li, N, Demarco, V, and Neu, J. 2004. *Pediatr Res* 56:250–255.)

- **Potential treatments:** Different measures may be taken to promote delivery, including physical removal of pups lodged in the birth canal, which can be done manually with lubrication and gentle retraction using forceps. Provision of supportive sterile fluids (1–3 ml SC of warmed 0.9% NaCl or lactated Ringer's solution [LRS]), calcium gluconate salt to increase contraction strength (100 mg/kg IP given 10 to15 min before oxytocin), with subsequent oxytocin (0.1–1.0 unit SC) may help stimulate delivery and potentially improve the health of the mother (Narver, 2012). It should be clarified that little scientific evidence exists demonstrating the benefits of oxytocin, a strong uterotonic drug, to dystocic mice (Narver, 2012, Schowalter et al., 2011); however, anecdotal reports indicated it is administered frequently to dams in dystocia. Administration of oxytocin may have adverse and confounding effects on research, especially for behavior studies, and administration may be contraindicated (Narver, 2012).

Despite the known critical role of prostaglandins in regulating murine parturition, prostaglandin therapy (2.5 μg prostaglandin F2α [PGF2α] SC) has not been shown to be more effective than oxytocin for alleviating dystocia (Chan and Washington, 2011). Analgesics, from drug classes other than nonsteroidal anti-inflammatory drugs (NSAIDs), may be administered in addition to other supportive care measures mentioned. There is concern that administration of NSAIDs, like ketoprofen, will inhibit cyclooxygenase, which is responsible for production of the prostaglandins essential to successful parturition in mice.

Promotion of environmental enhancements, including nesting material (cotton bedding or paper strips), minimizing disturbance to the cage, and judicious supportive care with fluids, caloric supplements, and palatable softened food, in addition to heating devices, may sometimes be sufficient to achieve vaginal delivery by animals in good condition (Narver, 2012). Any mouse in a compromised state during pup delivery should be frequently monitored, up to every 1 to 2 h, to determine if parturition is progressing and to assess health of the mother (Schowalter et al., 2011). Options for management approaches for dystocia (Figure 2.17) are provided (Narver, 2012).

For extremely valuable pups, the decision can be made to perform a cesarean section for salvage of pups remaining in

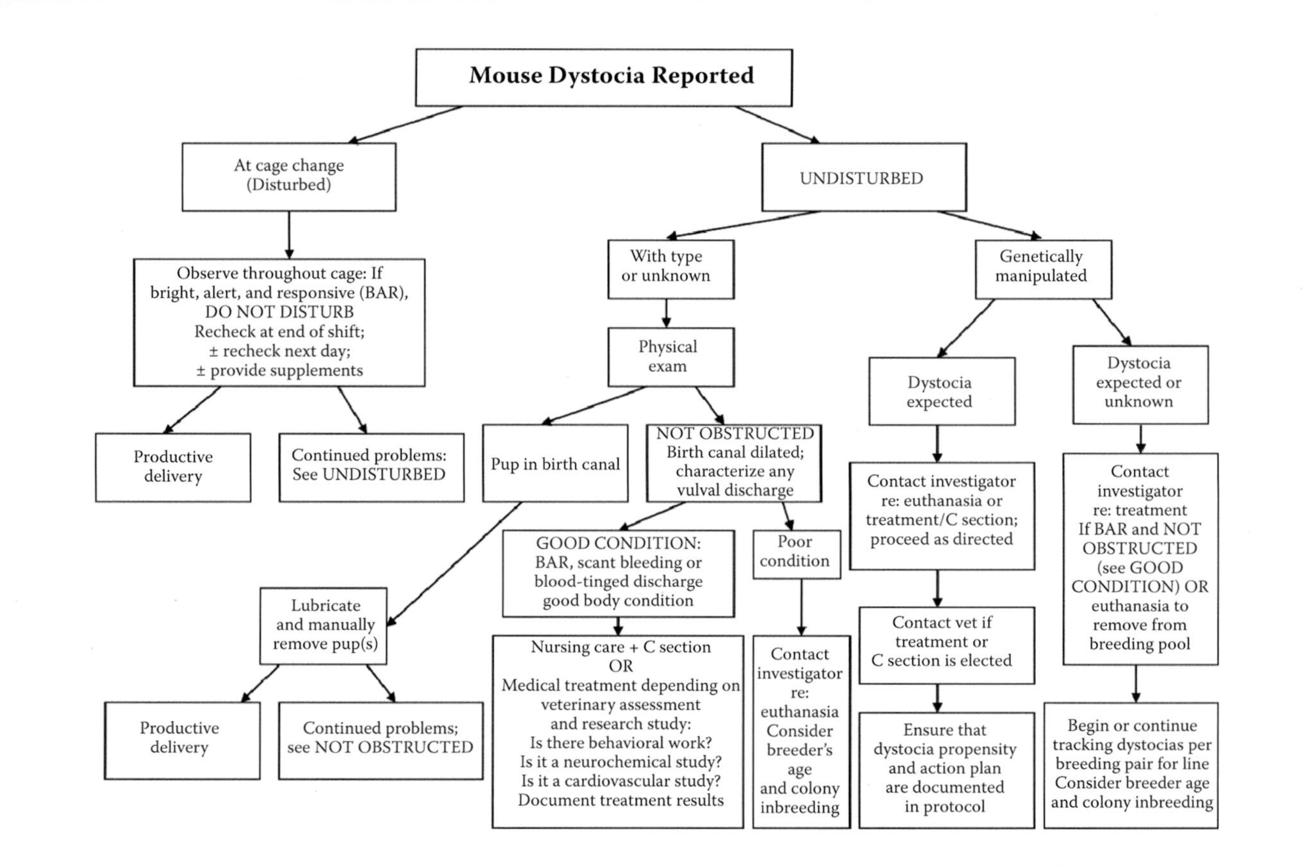

Fig. 2.17 Decision tree for dystocia management of mice. (Reprinted with permission from AALAS. Narver, HL. 2012. Oxytocin in the treatment of dystocia in mice. *J Am Assoc Lab Anim Sci* 51:10–17.)

the uterus; however, anecdotal reports indicated that these females do not return to normal breeding status if they survive the operative procedure, and that pups do not typically survive. If pups are collected in utero, they should be kept warm. It is imperative to stimulate activity by stroking pups gently with sterile soft materials (e.g., bandaging supplies or gauze) to stimulate responsiveness and breathing. Prior to placing pups with a foster dam, ensure that the animals are pink, taking breaths, and are responsive to stimuli.

Fight Wounds

- **Cause and impact:** Aggression in group-housed male laboratory mice is a widely recognized occurrence that can range from mild to severe as a clinical concern (Van Loo et al., 2003). In brief, male mice prefer social housing to individual housing; however, dominant males in a group-housed cage will show aggression toward subordinates. Aggressive behaviors have been linked to genetic background, odor cues, and the lack of an available "escape" from human handling or other mice within the cage. Injuries (Figure 2.18) are often targeted along the dorsal rump and tail area, as well as in the anogenital region.

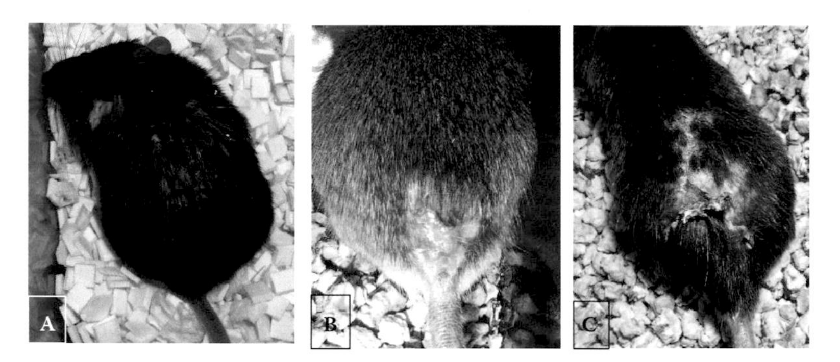

Fig. 2.18 Fight wounds affecting the dorsal aspect of various male mice. Wounds may be at the surface of the skin without readily apparent hair loss (A); wounding may appear like ulcerative dermatitis (B); healing of the lesions may occur by second intention, with sloughing of the haired scab over time (C). Ventral aspects of these animals should be physically examined to determine any further extent of injury. (Images courtesy of University of Pennsylvania, ULAR.)

- **Potential treatments:** Animals must be physically handled and thoroughly examined to palpate the extent of potential wound injuries both dorsally and ventrally.

  Cohoused males with fight wounds should be separated for the betterment of animal welfare, with the dominant male (most often the sole mouse in the cage with no apparent wounds) moved to an individual cage when aggression reaches unacceptable levels and group-housed males have incurred wounding.

  Individual housing is recommended for highly aggressive strains such as Swiss/CD-1 and FVB; however, the sheer impact of isolation from conspecifics may lead to increased aggression, and the addition of enrichment devices to the cage may be of benefit. Nesting material is the enrichment device of choice for group-housed male mice. For group-housed mice with the dominant animal removed, transference of used nesting material from the old housing cage to the new cage with the dominant male can further reduce aggression. Interestingly, it has been suggested that housing male mice in groups of three diminishes aggression (compared to groups of five or eight), indicating that the dominance hierarchy is more stable in smaller groups (Van Loo et al., 2000).

  With topical treatments similar to those applied for ulcerative dermatitis (UD; see relevant section on this topic), systemic analgesics (meloxicam 5 mg/kg daily for 3 to 5 days, administered SC or by mouth [PO]), and separation from aggressive mice, injured males very often recover to good health, with scabbing and eventual sloughing of the wounded skin.

  Note: Males with severe wounds may need to be euthanized, particularly if the anogenital region is scarred to a degree that the animal cannot urinate due to injuries surrounding the urethral opening and obstruction of urine outflow.

Fractures/Orthopedic Problems

- **Cause and impact:** Traumatic injuries related to fighting, improper handling and procedures, entrapment in caging equipment, and nutritional deficiencies may lead to broken bones (fractures) and related orthopedic concerns. Affected mice may present with a limp or lameness or other type of gait abnormality. A broken tail may appear kinked, with little evidence of swelling or bruising.

- **Potential treatments:** Depending on site of fracture (tail, limb), animals should be placed under frequent observation, assessed for ambulation, provided with pain medication (at least once daily for 3 days after the presumed fracture occurs), and possibly imaged to identify the site of the lesion by dual-energy X-ray absorptiometry (DXA) scanning or microCT (Figure 2.19).

Rodents typically can ambulate well despite tail and limb fractures and may not show overt evidence of distress. In time, fractures should heal on their own, depending on location and severity of the breakage. Bandages can be placed to immobilize fractures; however, restraint collars (see Chapter 4, "Restraint Collar Considerations") may need to be placed to prevent bandage removal (Hawkins and Graham, 2007).

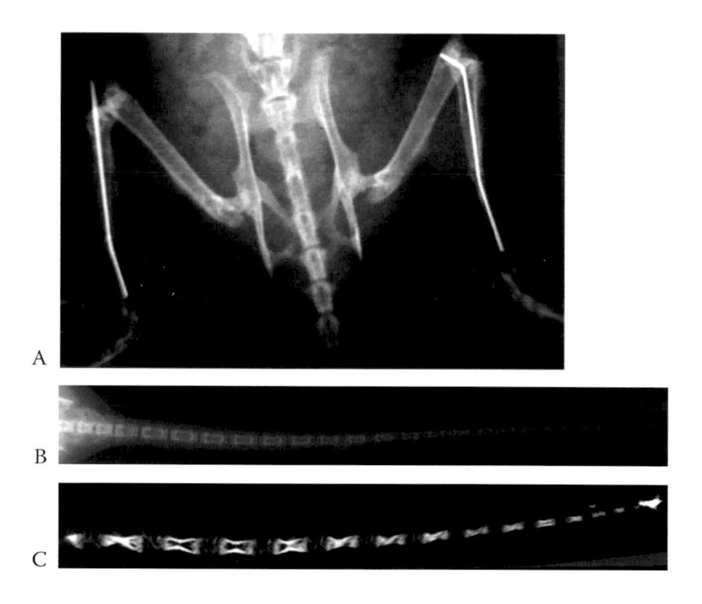

Fig. 2.19 Fracture models of murine tibial injury and repair by pinning may be imaged using DXA scans (A). This imaging technology can assess tail health and bone density (B), but greater distinction of vertebrae is capable using microcomputed tomography (C). (Images courtesy of University of Pennsylvania; K. Hankenson.)

If the animal is unable to ambulate or typical mobility is compromised, prompt euthanasia should be considered for animal welfare.

Hemorrhage

- **Cause and impact:** Hemorrhage (active bleeding) may be secondary to a number of physiological abnormalities, including trauma, thrombocytopenia, or experimental treatments. Certain mouse models of hemophilia (see relevant section on this topic) may exhibit this clinical symptom following routine procedures, like blood sampling. As well, bleeding may be due to lacerations, secondary to fighting, or because of improper hemostasis following tail clipping for genotyping.
- **Potential treatments:** Evidence of blood on any animal or in the housing cage should require immediate attention to ascertain the source and potentially to provide hemostasis to stop continued blood loss. Depending on the degree and source of blood, one can apply direct pressure to the site or styptic powder. Silver nitrate sticks are not recommended as they tend to be an irritant and leave a persistent chemical "burn" on the skin following use. Cautery applied using cordless disposable high-temperature loop tips (e.g., MediChoice®) work well for small lesions as long as the animals are under anesthesia at the time of application. If assisted wound closure is necessary, it is recommended to use stainless steel staples or tissue glue as routine sutures may be chewed out, and bandages may be poorly tolerated. Consider application of an appropriate size restraint collar on mice to prohibit oral access to suture sites (see Chapter 4, "Restraint Collar Considerations").

Moribund, Weak, or Paralyzed Condition

- **Causes and impact:** Hind limb weakness (paresis) and paralysis in laboratory mice may be associated with trauma, dysfunction, and weakness of the musculoskeletal and nervous systems, as well as infections caused by undesirable colony pathogens that may particularly affect immunodeficient strains. These animals may present with abnormal gait, ataxia, or dragging one or more limbs during ambulation.

Fig. 2.20 An adult female BALB/c mouse, 12 days postinoculation with herpesvirus, demonstrated neurological deficits, including tremors and an inability to grip and place feet normally. The animal was treated with supportive fluids and supplemental gel diet; the experimental protocol did not permit the administration of anti-inflammatory or analgesic medications due to interference with research outcomes.

Additional causes of weakness and paralysis may include inherited neurologic diseases, like myelin disorders or neuronal degeneration. Clinical signs develop in young mice carrying certain recessive mutations (i.e., *jp/Y, shi, mnd, wst*). Neoplasia and nonneoplastic diseases, such as osteoarthritis, bone fractures, or peripheral neuropathies (Figure 2.20), may also occur, particularly with increasing age of animals (Ceccarelli and Rozengurt, 2002).

Models of spinal muscular atrophy (SMA), a neurodegenerative disease of human children, have been established, with SMA mice dying by 2 weeks of age if untreated. Mortality has been linked to the phenotype of muscle weakness, but secondarily to malnourishment as the affected pups are outcompeted for access to nursing during the preweanling phase (Narver, 2011).

Moribund animals are typically those that are alive but nonresponsive to gentle manipulation by personnel, and they tend to be isolated from cage mates. This state may be expected for certain experimental models, but typically is irreversible despite concerted efforts to administer supportive care.

- **Potential treatments:** Institutions and laboratories need to ensure screening of biological materials prior to injection into colony animals to ensure cells and murine-derived injectables are pathogen free for agents, including MHV, mouse encephalomyelitis virus, and lactate dehydrogenase-elevating virus (LDV).

Certain infectious disease models may induce a moribund state from which animals can recover. It is critical to increase the frequency of monitoring and determine humane endpoints that eliminate prolonged suffering (see Chapter 4, "Humane or 'Clinical' Endpoint Considerations" and "Experimental Autoimmune Encephalomyelitis and Demyelinating Disease Model Considerations"). Animals that are paralyzed will need to have their bladders expressed two to three times daily, and topical lanolin ointment (Lansinoh®) should be placed on the ventral abdomen and hind limbs to prevent or minimize urine scalding of the skin.

Affected animals can be provided with fluid administered subcutaneously and offered nutritional support at the level of the cage floor (see Chapter 4. "Nutritional Therapy Considerations" and "Fluid Therapy Considerations"). Softened bedding substrates can be provided for comfort, and particularly weakened or moribund animals should be provided with heat and potentially separated to avoid further injury from cage mates.

Mouse models, like that for SMA, can also be provided with hand feeding to support nutritional requirements; antibiotic and analgesic medication may also be provided as needed (Wagner et al., 2011). Provision of ad libitum nutritional gel supplements on the cage floor has been shown to improve caloric intake and promote survival in neurodegenerative mouse models with persistent tremors (Black et al., 2011).

More often than not, moribund and paralyzed animals will require euthanasia if there is no improvement or change in activity status within 24 h of initial presentation.

Mortality (Sudden Death)

- **Cause and impact:** Sudden death in mice is a common occurrence, often without any premonitory signs, or may be secondary to conditions like acute toxicity, degenerative disease, seizures, subclinical infection, or vascular dysfunctions that were unrecognized prior to death.

There may be strain-related conditions of which to be aware, such as spontaneous death in apparently clinically healthy FVB/n mice with no premonitory symptoms at around 4 months of age. These mice have been described to have wet fur below the mandible and on the ventrum of their necks at the time of death. Histology on these animals identified multifocal areas of neuronal necrosis and loss in the cerebrum, which has been linked to the "space cadet syndrome" described for FVB/NCR mice. While both genders are affected, female FVB are more predisposed to sudden death. The condition is thought to result from neuronal necrosis in the brain due to seizure activity. Incidence of this condition may be underdiagnosed and should be considered in evaluation of FVB/n wild-type and transgenic phenotypes (Rosenbaum et al., 2007).

- **Potential treatments:** While there is no treatment for animals that succumb to sudden death, the opportunity should not be overlooked to necropsy animals that have recently died (within 4 h) and perform histopathology on fresh tissues to attempt to determine the root cause of death.

Ocular Lesions

- **Cause and impact:** Ocular abnormalities (Figure 2.21) are frequently identified in laboratory mice and may appear without any obvious etiology. Eyelids may be squinted closed over the eye; the globe itself may have alterations (ulcers) or opacities (cataracts); there may be discharge noted; or an altered size and shape of the eyeball may lead to exophthalmos (forward projection of the globe out of the socket). Any abnormal swelling or mass development in or around the eye should be reported. Animals will often appear to be otherwise behaviorally normal despite the ocular lesion.

  It is important to be aware that many common laboratory mouse strains (e.g., C3H, FVB/N, SJL/J, SWR, and some outbred Swiss mice) are blind due to genomic mutations (Danneman et al., 2012). Mice with microphthalmia often have abnormalities in a variety of ocular structures; this condition is common in C57BL/6 and related mice, with increased incidence in females compared to males.

  Additional things to rule out should include strain-related disease, glaucoma, congenital abnormalities, trauma

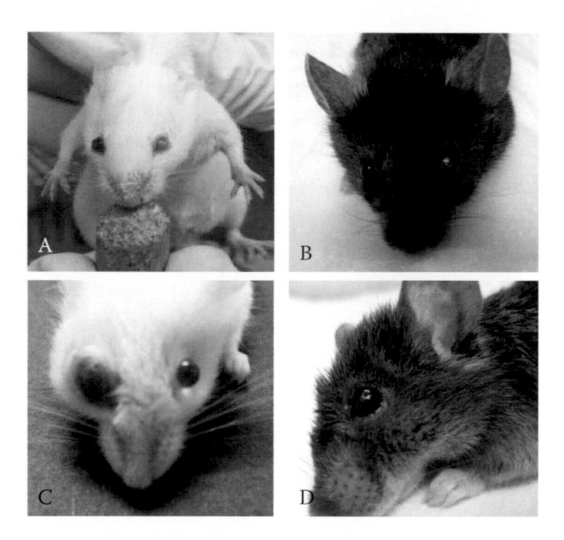

Fig. 2.21 Representative ocular lesions that should be reported as clinical cases for monitoring of progression and management of pain: (A) cloudy left eye, (B) bulging or proptosis of the left eye, (C) retro-orbital neoplasia of the right eye, and (D) corneal pitting and ulceration of the left eye, likely secondary to failure to apply ophthalmic ointment while under anesthesia. (Images courtesy of the University of Pennsylvania, ULAR.)

(perhaps secondary to a recent retro-orbital bleed), compound administration, light sensitivity due to any expected neurological disorder, retro-orbital abscessation, or neoplasia. Acute reversible corneal lesions have been documented in mice, attributable to a side effect of xylazine for anesthesia (Calderone et al., 1986).

- **Potential treatments:** Certain ocular lesions can be avoided through the routine use of eye lubrication ointment (e.g., Puralube™ or Rugby® Sterile Artificial Tears Ointment Lubricant–Ophthalmic Ointment) for any mouse undergoing anesthesia for any procedure. This avoids desiccation of the eyeball that has been otherwise shown to lead to corneal opacities.

  Routine prophylactic cleaning of the conjunctiva with sterile swabs and saline will assist with removal of debris. Fluorescein stain can be applied to check for corneal ulcers and abrasions. Prior to staining, proparacaine (0.5%; 1 to 2 drops per eye) may be applied topically directly to the globe for anesthesia.

Application of topical ophthalmic ointment, with or without added antibiotics, is warranted as a first-line approach to an eye injury. Certain lesions will be painful, with notation of animals scratching at the eye and face; these animals should receive topical anesthetic drops (proparacaine 0.5%) and systemic analgesics (meloxicam 5 mg/kg SC) for pain relief.

If an animal presents with proptosis, the lid margins around the globe should be retracted gently following cleaning of the eye surface and provision of topical ointment for lubrication. Gentle pressure should be applied to the intact globe to reduce the prolapse, and topical wetting drops and ophthalmic antibiotics can be provided for 7 to 10 days after replacement. Topical steroids should be avoided, but NSAIDs appropriate for ophthalmic application can be considered (Hawkins and Graham, 2007).

Surgical enucleation can be performed under anesthesia for critical injuries, including ulcerations (Cote et al., 2011).

Perineal Swelling

- **Cause and impact:** Perineal swelling presents as an enlargement of subcutaneous tissues on the ventral and distal abdomen in both males and females and is typically benign (Figure 2.22). Male mice in particular may be reported for

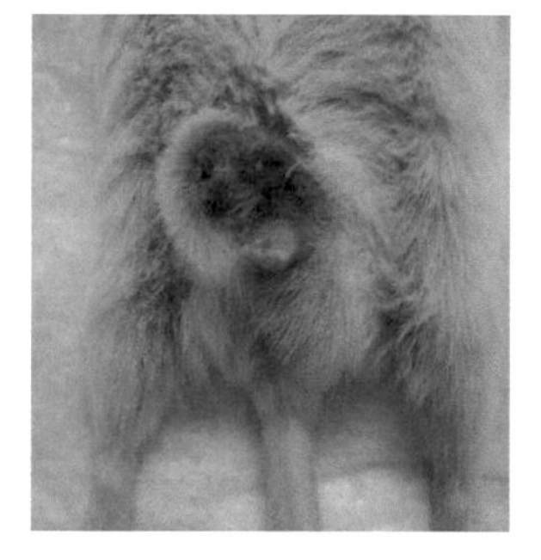

Fig. 2.22 Representative perineal swellings that should be reported as clinical cases for monitoring of progression, potential urinary obstruction, and management of pain: female mouse with an abscessation of the clitoral gland (left); male mouse with abscessation of the preputial gland (right). (Images courtesy of the University of Pennsylvania, ULAR.)

perineal swelling, which is believed to have a genetic component, and there may be translocation of abdominal organs into the perineal space or cysts within the bulbourethral glands (Hill et al., 2002). Preputial gland abscesses secondary to bacterial infection have long been recognized as a clinical issue in laboratory mice, and often *Staphylococcus* or *Pseudomonas* are cultured (Hong and Ediger, 1978). Bacterial agents may enter through the urethra or potentially through fight wounds in males. Male mice have also been found to have cholesterol granulomas with hemorrhage, in conjunction with bulbourethral dilations (Dardenne et al., 2011). Females have been reported to have clitoral gland abscessation (Alworth and Nagy, 2009) and swellings that were diagnosed subsequently as mammary adenocarcinomas, likely strain related in origin (Naff et al., 2005).

It is important to train animal care and investigative staff to recognize these, and other, reproductive anomalies in breeding colony mice. Because of the swollen perineum, mice with imperforate vaginas are often mistaken as males. The presenting complaint may be failure of a breeding unit to produce pups, when the issue is actually that a normal female has been paired with a second female with an imperforate vagina.

- **Potential treatments:** Many of these perineal swellings tend to be incidental findings in mice that are otherwise bright, alert, responsive, and in good body condition. Fine-needle aspirates can be performed on the swellings to ascertain if they contain purulent material, as many are caused by secondary bacterial infections. While research data can be obtained from animals with perineal alterations, if a genetic component is suspected in affected animals, and if these mice are in breeding colonies, it may be of benefit to cull affected mice to avoid perpetuation of the condition in offspring (Rubino et al., 2004).

  If the swelling is due to imperforate vagina in the female, the condition can be treated with surgery, yet with a limited chance that the female will return to reproductive function following surgery (Ginty and Hoogstraten-Miller, 2008):

  - The vaginal membrane can be transected to release the retained mucoid debris, and the newly formed orifice can be further enlarged using blunt scissors to ensure an adequate opening and good drainage. Take precaution when flushing the vagina and uterus as these thin, dilated, and flaccid tissues may be damaged.

- Supportive care involves instillation of triple antibiotic ointment with hydrocortisone into the vaginal orifice, followed by 1 to 2 drops of a local anesthetic agent (e.g., bupivacaine) along the incision site for 3 days.
- Antibiotics (trimethoprim-sulfa SC) can be administered twice daily in 1 ml 0.9% saline for up to 7 days.

Realistically, the percentage of mice that return to breeding soundness postsurgery is low, potentially due to the presence of other reproductive abnormalities or to permanent damage caused by the extreme distension of the uterus. Therefore, surgical repair may be primarily incorporated as a salvage procedure so that valuable females can be used for other experimental manipulations.

Poor Body Condition

- **Cause and impact:** Mice may present with thin, hunched, and ruffled appearance (Figure 2.23) without much forewarning; this may be related to a wide range of factors involving anatomy and physiology, strain phenotype, and experimental manipulations. Mice may have teeth that have overgrown due to malocclusion and thus be unable to prehend and ingest food. As well, immunodeficient strains may have underlying (subclinical) infectious disease that should be considered as a differential and may result in isolation of these animals from the colony. Animals may have ruffled fur and hair loss due to inflammation, autoimmune responses, pruritis and subsequent scratching, or due to ill health, such that the mouse is not attending to self-grooming habits.
- **Potential treatments:** The logical and prioritized causes should be treated first, typically including administration of a bolus of fluids subcutaneously, nutritional support, and heat supplementation. Dental assessments should be conducted to determine if there is a loss of alignment of incisors (malocclusion); if so, these teeth can be trimmed with nail scissors and should be monitored for future overgrowth. Softened food and supplements can be provided on the cage floor following trimming.

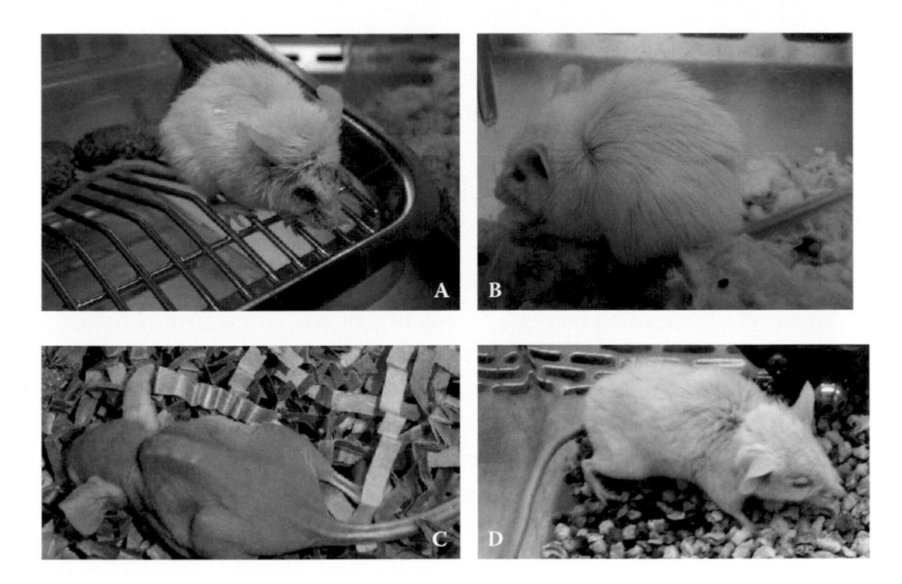

Fig. 2.23 Mice can present with overall poor body condition, identified by lack of normal activity, ruffled fur (A, B, and D); hunched or abnormal postures (A–D); or thin appearance in the range of a 1–2 on the BCS scale (A, C, and D). Often, animals in poor condition are enrolled in experiments intended to cause an adverse outcome, or the condition may be related to inherent immunosuppression or underlying spontaneous disease, like tumorigenesis. A thorough history and experimental description will be essential to compile the necessary database of information to formulate a treatment plan. (Images courtesy of the University of Pennsylvania, ULAR.)

Blood samples can be collected to assess serum chemistry and complete blood counts, as well as provide serum to conduct a pathogen screen (Wiedmeyer et al., 2007). Ectoparasites should be ruled out through skin scrapes, and fungal cultures should be collected to discern if this might be contributing to hair loss.

If the animal is not eating a provided experimental source of food or pellet or if the treatments are potentially toxic or unpalatable, this should be further discussed with the research team to ensure the animal is meeting its caloric needs.

If animals were expected to succumb to experimental disease, the treatment efforts should be aimed at assisting with comforting the animal in an attempt to achieve the experimental endpoint. BCS should be assessed daily and

BWs checked routinely to track any further losses. Animals that become quiet, less alert, and unresponsive should be considered for euthanasia prior to reaching a moribund state and spontaneous death.

Rectal Prolapse

- **Cause and impact:** An eversion of the rectal mucosa beyond the rectal opening (Figure 2.24) is not an uncommon finding in laboratory mice and may range from mild to severe enough to warrant euthanasia. This prolapse may be due to strain-related phenotypes, the efforts of parturition, or intestinal infection (e.g., *Helicobacter* spp.) or other conditions that cause diarrhea or straining to defecate. The mucosa may remain moist and the prolapse actually identified due to adherence of bedding substrate to the rectum in an animal that otherwise has a normal body condition and activity level.
- **Potential treatments:** Husbandry management would indicate that the bedding substrates should be changed to softened paper materials for animals with rectal prolapses. Any adhered bedding substrate should be removed from the mucosal tissue to determine the severity and state of the

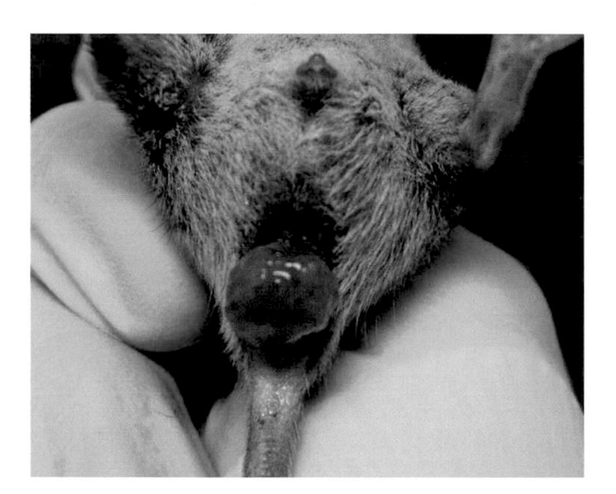

Fig. 2.24 Rectal prolapse of moderate severity; the mucosa is dark red with pinpoint areas of hemorrhage. This mouse would need further assessments to ensure normal defecation can occur, and topical ointment should be applied to keep the tissue moistened. Changing the bedding to a soft substrate could prevent further irritation of the tissue. (Image courtesy of the University of Michigan, ULAM.)

prolapse, and moisturizing treatment with triple antibiotic ointment or lanolin can be applied to dry and desiccated tissues.

If pathogen infection is suspected (e.g., pinworms, *Helicobacter*, or *Citrobacter*), treatment with antiparasitics or antibacterial therapy appropriate to the pathogen should be instituted promptly.

Successful correction of rectal prolapse has been described using a technique that permits the repositioning of the rectum inside the anal cavity, thus preventing tissue cyanosis and necrosis (Koch, 2010). Topical triple antibiotic ointment with hydrocortisone is first applied to the perineal area to reduce inflammation. The rectum is then pushed back inside the body using a blunt probe, then a small amount of tissue glue is applied on the borders of the anus to temporarily maintain the rectum in place and prevent it from prolapsing when the mouse is passing feces. Daily application of antibiotic ointment topically at the site is recommended, along with daily administration of an injectable NSAID (e.g., carprofen or meloxicam 5 mg/kg SC) for 3 consecutive days if this analgesic choice does not interfere with the research outcomes (Koch, 2010).

If rectal prolapses are prevalent in particular strains, scoring systems for severity of rectal prolapse to determine points for humane intervention may be beneficial (Figure 2.25).

Respiratory Distress

- **Cause and impact:** The respiratory system for mice has been reviewed (Kling, 2011), and adverse clinical signs dependent on some aspect of the respiratory system can include nasal discharge, ocular discharge, sneezing, audible "chattering," dyspnea, open-mouth breathing, cyanosis, and head tilt.

As mice are obligate nasal breathers, the development of respiratory distress in the face of infectious pulmonary disease may be rapid. Infectious agents that may be implicated include viruses (e.g., Sendai virus) and bacteria (e.g., *Mycoplasma*). If the animals are deemed to be in stable condition (not overly stressed), one may consider using imaging methods of radiography to ascertain if there are consolidations, opacities, or other abnormalities in the lung field that are contributing to the condition.

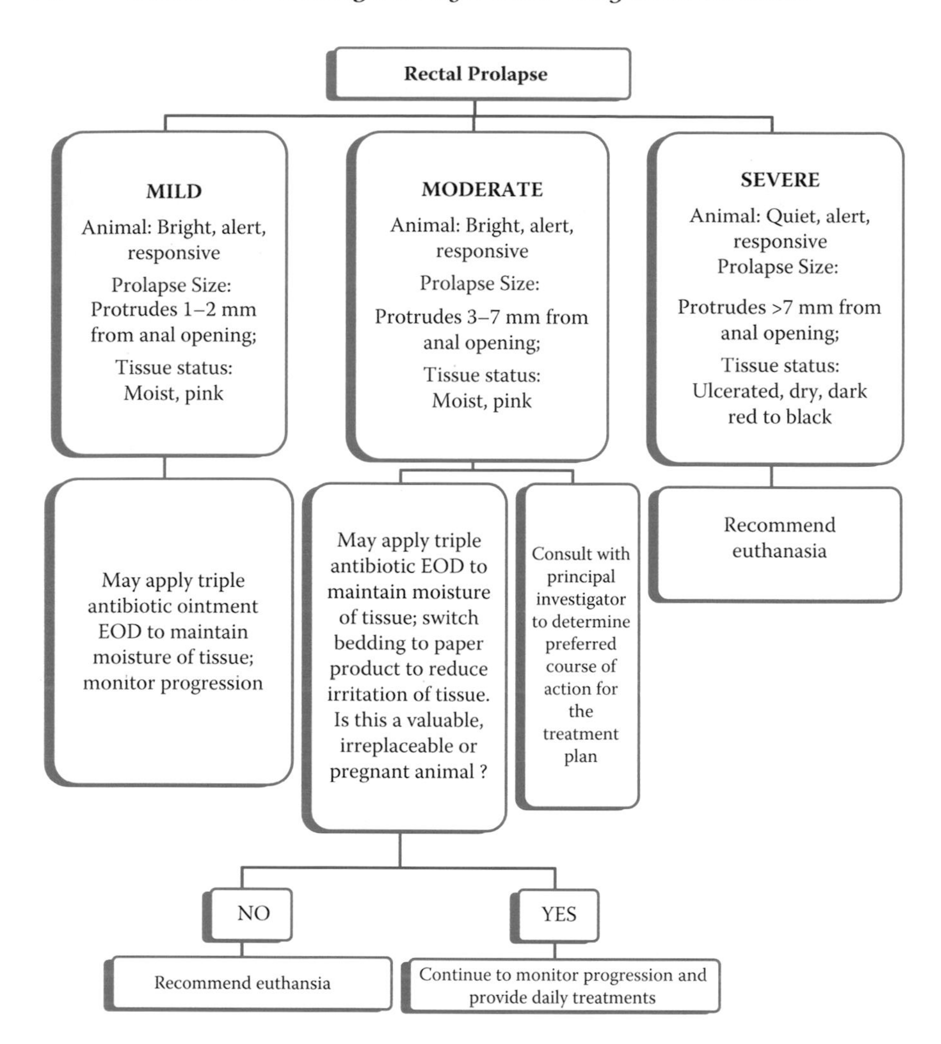

Fig. 2.25 Decision tree for management of rectal prolapse in rodents. (Modified from University of Washington, DCM.)

- **Potential treatments:** Respiratory distress in any animal warrants prompt administration of supplemental oxygen; it is recommended to place rodents in a small induction chamber or place the entire animal inside a large anesthetic face mask to deliver oxygen rapidly. The chamber can be kept somewhat cool relative to ambient temperatures as animals in distress may have increased core temperatures (hyperthermia) due to their respiratory efforts. Rodents should be observed closely for improved breathing patterns and changes in mucous membrane color toward a pink/red hue.

If an undesirable infectious pathogen is suspected as the root cause, it is recommended to quarantine the affected animal and potentially perform euthanasia, tissue harvest, and serology to diagnose the agent. These diagnostics may be critical if there is concern of pathogen spread into additional mouse colonies in the facility.

For certain animals, depending on the experimental plan, it may be of use to provide light sedation to calm the respiratory effort and diminish anxiety. It may be appropriate to administer diazepam (1–3 mg/kg IP) or midazolam (0.5–1.0 mg/kg IP) as described (Oglesbee, 2011). If a bacterial cause is confirmed, antibiotic sensitivity testing will direct selection of appropriate systemic antibiotic treatments.

Seizures

- **Cause and impact:** Generalized seizures are caused by paroxysmal cerebral dysrhythmias and are characterized by loss of consciousness, muscle contraction (tonus), and jerking (clonus) (Aldrich, 2005). Seizures may present in animals with a sudden onset of shaking or chewing; with circling, momentary paralysis or "freezing"; or with convulsions (Figure 2.26).

  Episodes typically are brief and may arise spontaneously after handling of rodents; they can be accompanied by autonomic dysfunction (urination/defecation). In status epilepticus, seizures occur in rapid succession without recovery between them; this intense neuronal activity can cause metabolic derangements with damage to neurons and brain swelling. The exertional muscular activity during seizures can predispose the rodent to hyperthermia, hypercapnia, hypoxemia, and metabolic acidosis (Vernau and LeCouteur, 2009).

  Models of epilepsy (Fisher, 1989) may be desired for certain experimental protocols, and seizures may be linked to the transgenic or knockout strain or may be due to development of a deleterious mutation (Pesapane and Good, 2009).

  Seizures in FVB mice have been described (Goelz et al., 1998); observations of seizure activity were made of mice while in their cages, when handled for tail tattooing and fur clipping, as well as during facility fire alarms. The majority of affected animals were female FVB/N. Clinical presentations included facial grimace, chewing automatism, ptyalism

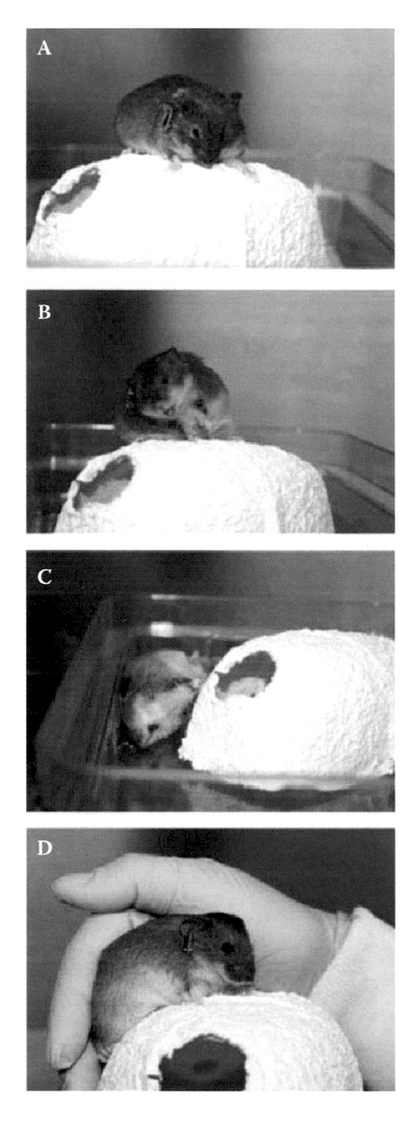

Fig. 2.26 Sequence of events leading to full seizure in an affected mouse: (A) After handling, the mouse is placed on the top of a Shepherd Shack® (Shepherd Specialty Papers, Portage, MI) and starts to experience paralysis. (B) Paralysis worsens, and the animal's mouth opens; the mouse twists and shakes. (C) The animal falls from the Shepherd Shack and is convulsing with its mouth open. (D) The animal is fully recovered after approximately 5 min. (Reprinted by permission from Macmillan Publishers Limited. Pesapane, R, and Good, DJ. 2009. Seizures in a colony of genetically obese mice. *Lab Anim (NY)* 38:81–83.)

with matting of the fur around the neck and on forelimbs, and clonic convulsions that, at times, progressed to tonic convulsions and death. In some mice, only nonspecific signs of disease were noted, such as lethargy, moribund state, and matting of the fur (from hypersalivation); blood glucose values remained within normal limits.

Animals may undergo multiple unobserved seizure episodes, culminating in a terminal lethal convulsion; these animals will likely be found dead yet appear in good body condition without other evidence of an overt cause of death. In keeping with reports of finding spontaneously dead FVB/N mice, it has been documented that susceptibility to seizure activity increases during the dark cycle, when the laboratory rodents in a typical research facility would not be routinely observed by personnel who could otherwise track and report on seizure activity (Goelz et al., 1998).

Generalized spontaneous seizure disorders in mice have been documented in dilute lethal (*dl*), quaking (*qk*), and wobbler-lethal (*wl*) strains. Certain inbred strains, including DBA/2J, SJL/J, and LP, may have generalized lethal seizures when exposed to auditory stimuli (Fuller and Sjursen, 1967). Audiogenic seizure susceptibility varies widely between strains and appears to be influenced by genetic and environmental factors (Goelz et al., 1998).

- **Potential treatments:** Neurologic examinations in rodents are not typically easy to perform and may be based on observations of return to clinically normal behaviors after a seizure, including returns to expected eating and drinking patterns and activity. It is important to note that animals may be neurologically abnormal for days after a seizure (Vernau and LeCouteur, 2009), which can have impact on the collection of accurate experimental data and the overall model under investigation. As well, once an animal has seizured, it may be indistinguishable from normal cage mates within a period of a few hours. Questions to consider when trying to determine the history of the seizures (particularly if noted in a colony of genetically similar animals) include the following (Pesapane and Good, 2009):
  - Is this condition linked to a gene of interest in this line of mice?
  - Is it correlated with gender or age at onset?

- Could there be an environmental or infectious cause?
- Is this condition heritable?

Emergence of seizures in a colony that otherwise has not previously been reported should instigate a review of breeding record keeping to determine the frequency of occurrence and to determine if this trait originated with a single breeding pair linked to a particular cohort of offspring maintained within the colony.

Drug treatments are not typically described in mice, although diazepam is a first-line agent of treatment in other species (Vernau and LeCouteur, 2009). If seizures are unresponsive to benzodiazepines, one can attempt treatment with phenobarbital (4 mg/kg intramuscularly [IM], give twice at 20 min apart; continue this 12 h later at 2 mg/kg PO). Dextrose may also be administered if hypoglycemia is present postseizure (Hawkins et al., 2007).

Trauma

- **Cause and impact:** As mentioned in other sections, traumatic injuries in routinely housed laboratory mice may be caused by improper handling or intracage fighting, improper administration of experimental agents (e.g., tumor cells injected extravascularly), or entanglement in caging equipment (Figure 2.27).

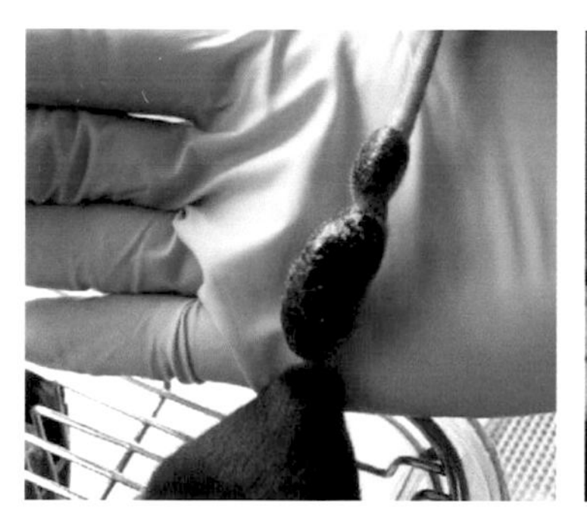
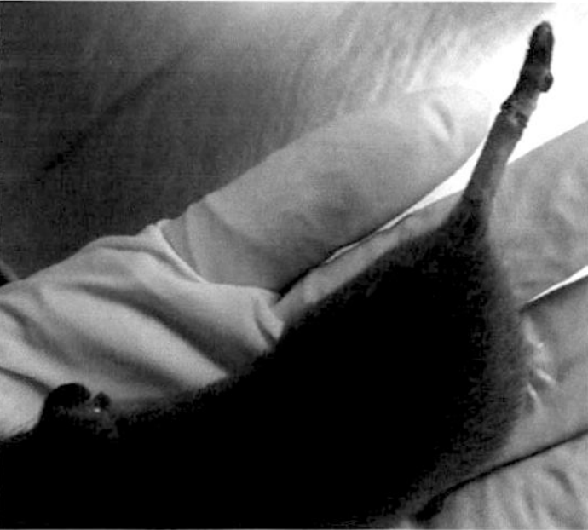

Fig. 2.27 Trauma to the tail caused by inappropriate tail vein injections of tumor cells (left) and by damage from cage mates following injury from cage materials (right). These animals should receive pain management and potentially surgical tail amputation to remove the affected necrotic tissues. (Images courtesy of University of Pennsylvania, ULAR.)

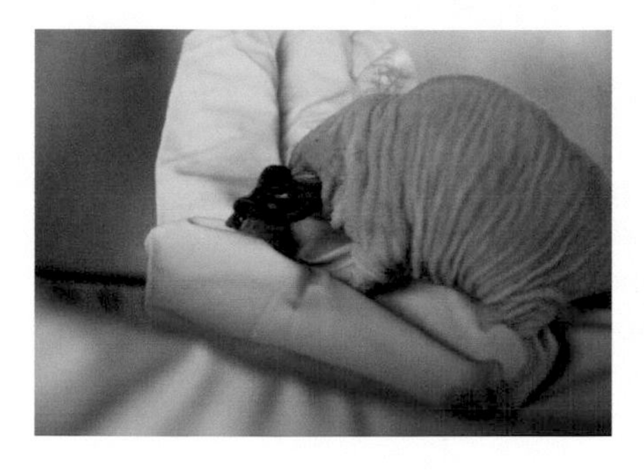

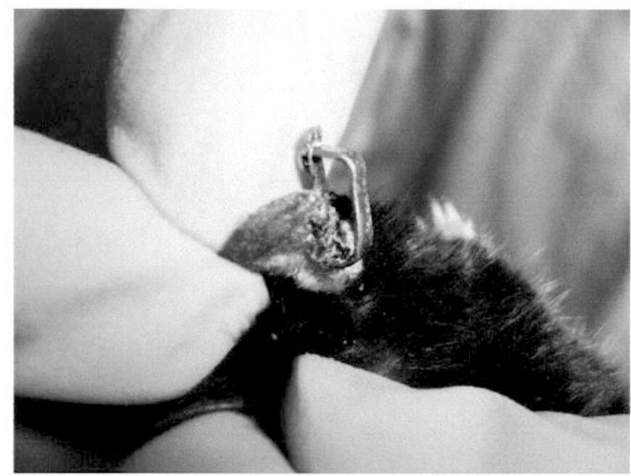

Fig. 2.28 Trauma to the ears related to metal ear tag placement. A nude mouse ear became entrapped in cage materials, resulting in avulsion and necrosis of the tissue, which required amputation of the ear (left); foreign body reaction to the metal tag may result in self-injury and scarring (right). It is recommended that the ear tags be removed if an allergic response is suspected. (Images courtesy of University of Pennsylvania, ULAR.)

Trauma may have an impact on any part of the body, including the face, limbs, paws, or ears, often due to metallic identification ear tags becoming stuck in wire bar feeders, j-feeders, and wire bar lids. These tag entrapments are precipitated by the routine exploratory and climbing activities in which mice engage as part of their species-specific behavior (Figure 2.28).

Animals may also self-injure, related to their attempt to ameliorate painful sensations, and cause severe self-trauma to eyes, to existing dermatitis, or to limbs that may have become injured secondary to environmental or experimental influences.

- **Potential treatments:** Animals should be rescued from any compromising entrapment in the cage as soon as it is noted and then further evaluated for injury and potential fractures to legs and tail. Any active hemorrhage should be immediately assessed for source of the bleeding, and manual hemostasis should be provided.

Ears may be avulsed and necrotic following entrapment; removal of pinnal tissue can be done with pen-like cautery tools to provide a clean dissection of the affected tissue away from healthy tissue. Animals should be under systemic anesthesia prior to the cautery recommended for amputation of ear tissue.

Animals that potentially drop a distance when being handled by personnel should be recaptured and examined for injury; in most contemporary housing facilities, it is typically not acceptable to place an animal that has fallen onto the floor or out of the flow hood or biosafety cabinet back into the original cage due to biosecurity concerns. Instead, these animals should be singly housed (quarantined) for observation and isolated due to their potential contamination from exposure to macroenvironmental elements.

Ulcerative Dermatitis

- **Cause and impact:** In the research animal community, there are diverse anecdotes and hypotheses about the etiology of ulcerative dermatitis (UD). In contemporary facilities, dermatitis is noted in mice from multiple background strains, not just those with a C57BL/6 background. Development of severe lesions can occur rapidly; areas of ulceration can expand, most often due to self-mutilation, in the course of less than 24 h. The more commonly noted sites of ulcerative dermatitis are between the scapulae and on both dorsal and ventral aspects of the neck area; however, facial dermatitis is not uncommon (Figure 2.29). A chronic nidus of inflammation, such as that occurring from oral mucosa pierced by shed hairs during grooming, has a significant link to the presence of ulcerative dermatitis (Duarte-Vogel and Lawson, 2011, Lawson, 2010).

  Ulcerative dermatitis has been reviewed comprehensively (Hampton et al., 2012, Kastenmayer et al., 2006), and the most important underlying factors have been documented to best synergize treatment modalities. The skin disorder has been attributed to some combination of infectious, genetic, behavioral, nutritional, immunological, endocrinological, environmental, and neurological factors. Neurological "skin-picking" disorders have been implicated as well. (Dufour et al., 2010).

  Histological assessments of UD lesions from C57BL/6 mice have indicated that the earliest detectable lesions involve follicular dystrophy, with degradation of the inner root sheath and defects in the hair fiber cuticle that may puncture the follicles and lead to inflammatory reactions (Taylor et al., 2005).

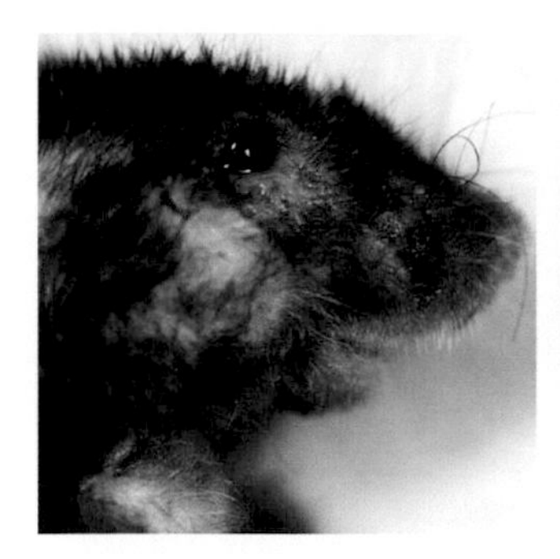

Fig. 2.29 Typical appearance of ulcerative dermatitis in mice. (Left) Mouse with facial dermatitis around the eyes, with scarring and retraction of the mandible due to lesions under the chin; nasal abrasions are likely secondary to excessive grooming activities. (Reprinted with permission from AALAS. Lawson, GW. 2010. *Comp Med* 60:200–204.) (Right) Mouse with severe ulcerative lesions and hair loss over the scapulae, affecting at least 10% of the body area. (Images courtesy of University of Michigan, ULAM.)

Progressive necrosing dermatitis of the pinna has been described in CD-1 outbred mice; initial lesions may be subtle and thus missed. However, peripheral necrosis of the pinna may develop with serous discharge, and the pruritus may cause the mice to self-injure and excoriate the area (Slattum et al., 1998).

- **Potential treatments:** Importantly, one should rule out other infestations, such as mites, that may lead to pruritis and ultimately dermatitis (Smith et al., 2003). Reports of attempted treatment (e.g., topically applied corticosteroids, antibiotics, or antiseptics) for pinnal necrosing dermatitis may not eliminate lesions but may help to control progression (Slattum et al., 1998).

It is critical in treatment of UD to understand that no application will likely eradicate all signs in all animals; however, even partial resolution or inhibition of further development can be an improvement to animal welfare and avoid the outcome of euthanasia prior to the animal reaching its anticipated experimental endpoint.

A comprehensive list of UD treatment types and options is provided next (in alphabetical order, depending on administration route, and with further specific instructions from available resources):

- *Topical application* (daily or EOD depending on severity of lesions):
  - **Antibiotics:** Selection should be based on culture and antibiotic sensitivity results of skin lesions; Neosporin or similar products are most commonly used as a front-line approach.
  - **Calamine liquid suspension with zinc oxide** (calamine lotion).
  - **Caladryl®** (camphor/pramoxine/zinc):
    - Main advantage of Caladryl is mitigating the scratching to allow lesions to dry and heal; improvement has been seen in up to 65% of cases (Crowley et al., 2008).
    - Clean affected skin gently with chlorhexidine solution with sterile gauze.
    - Apply Caladryl once daily in thin layer for 5 to 7 days using clean cotton swabs or clean gloved finger.
    - Healing ensues sequentially after cessation of scratching, with resolution of erythema and drying of lesions.
    - Treatment should continue after lesions have healcd grossly and should continue or be reinstituted if scratching is noted.
    - Treatment may be less successful in mice with multiple UD lesions or advanced full-skin-thickness lesions.
  - **Chlorhexidine** (0.2% solution):
    - Dilute 1 ml 2% chlorhexidine gluconate in a test tube containing 9 ml distilled water (1:10 dilution) and shake thoroughly.
    - Using sterile cotton swab or gauze pad, apply a generous amount to affected areas once daily for 7 to 14 days, followed by application of topical triple antibiotic ointment for severe cases (Lumpkins et al., 2006, Delgado et al., 2011).
    - Mice were noted to stop scratching within 3 days of this treatment, and most treated mice were lesion free for up to 7 months.

- **Cyclosporin** (0.2%) in a 2% lidocaine gel:
  - Apply mixture topically and supplement with 50 μg/ml gentamicin twice daily (Feldman et al., 2006).
- **Derm Caps®:** A proprietary combination of omega-3 and omega-6 fatty acids (contains safflower oil, a source of linoleic acid; borage seed oil, a source of gamma linolenic acid; fish oil, a source of eicosapentaenoic and docosahexaenoic acid; and vitamin E).
- **Diabetic foot ulcer medications:** May be beneficial due to inclusion of platelet-derived growth factors
- **EMLA®** (Eutectic Mixture of Local Anesthetics): A mixture of 2.5% prilocaine and 2.5% lidocaine cream that provides topical anesthesia with prolonged contact time (>15 min).
- **Gentian violet (GV):** An anti-infective ointment in an alcohol-based solution.
- **Green clay** (montmorillonite):
  - Clay is mixed with water to reach a thickened consistency and should be allowed to stand for 1 h prior to use.
  - Shave around the affected skin area (with mice under anesthesia) and then clean shaved area with dilute chlorhexidine and allow to air dry.
  - Apply clay poultice to a thickness of 5 mm and reapply every 3 to 4 days, as long as animal does not remove clay (Figure 2.30).
  - Improvement in UD was noted that ultimately resolved in 123 of 150 similarly treated mice (Martel and Careau, 2011).
- **Lansinoh:** Lanolin ointment used for keeping skin moist. It serves as a nontoxic emollient and is successful for treatment of dry skin for pups and nude mice (Taylor et al., 2006).
- **Natural unpasteurized honey** (Mathews and Binnington, 2002).
- **New Skin®:** Liquid bandage substance.
- **Panalog®:** Contains antibacterial, anti-inflammatory, and antifungal components.
- **Silvadene®:** A 1% silver sulfadiazine ointment.

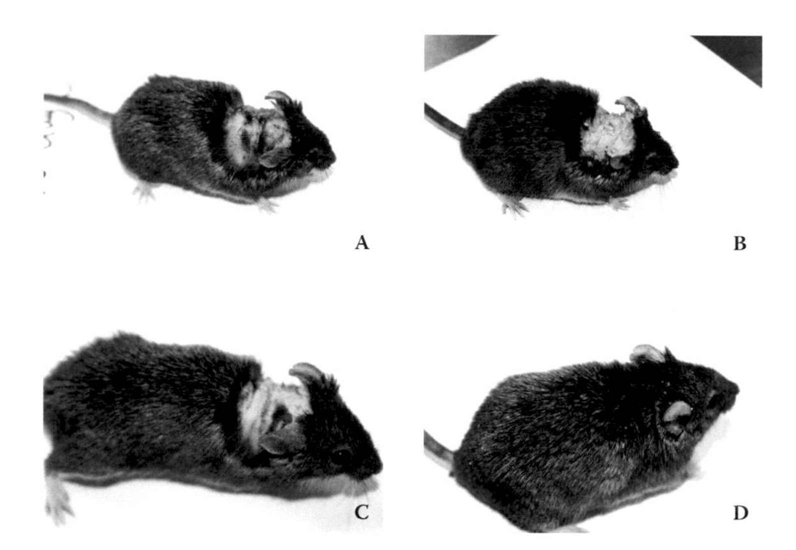

Fig. 2.30 Green clay therapy for mice with UD: (A) untreated mouse; (B) treatment of UD with topical green clay; (C) healing by day 4 postapplication; and (D) appearance of the mouse 3 months later. (Reprinted with permission from AALAS. Martel, N, and Careau, C. 2011. *Tech Talk* 16:2–3.)

- **Sterile saline (hypertonic):** To remove debris and bacteria.
  - Place 100 ml distilled water in screw-cap bottle.
  - Add 4.4 g of sodium chloride and shake well; loosely screw cap on bottle and steam sterilize bottle on liquid cycle in an autoclave.
  - Apply saline with sterile cotton ball; apply to crusts and ulcerative areas twice daily for 2 to 3 days.
  - Flush remainder of lesion using a sterile syringe or pipette; as healing progresses from granulation to epithelial tissue, treatment can be reduced to EOD (Mangete et al., 1992).
- **Steroids:** To combat inflammation.

• *Systemic:* May include antibiotics, anti-inflammatories, antihistamines, and analgesics, along with the following:
  - **Vitamin E**
    - When added to food (dosed at 1,600–3,400 IU/kg), has been shown to improve lesions greater than

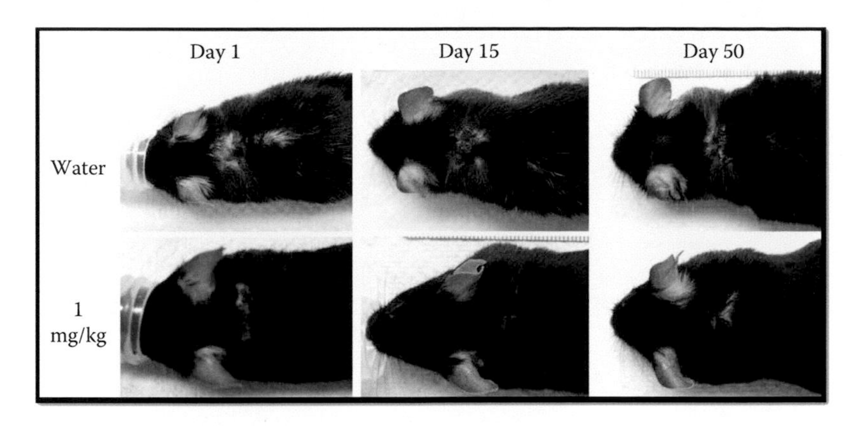

Fig. 2.31 Progression of healing UD lesions; animals on the bottom row were treated with systemic maropitant citrate (dosed at 1 mg/kg IP) compared to controls in the top row. (Reprinted with permission from AALAS. Williams-Fritze, MJ, Carlson Scholz, JA, Zeiss, C, Deng, Y, Wilson, SR, Franklin, R, and Smith, PC. 2011. *J Am Assoc Lab Anim Sci* 50:221–226.)

50% at the low-end dose (Barnard et al., 2006, Lawson et al., 2005); Derm Caps® can be the source of vitamin E, with 0.1 ml/day dripped onto a food pellet.

- **Maropitant citrate:** A neurokinin 1 (NK1) receptor antagonist.
  - For lesion improvement (Figure 2.31), dose once daily at 1 mg/kg IP for 5 to 10 days (Williams-Fritze et al., 2011).

- *Other miscellaneous adjunctive therapies:*
  - **Toenail clipping** (especially rear feet):
    - It is well documented that reduction in self-mutilation from scratching can diminish progression of dermatitis (Disselhorst et al., 2010, Martel and Careau, 2011, Mufford and Richardson, 2009).
    - To avoid scruffing the mice over the ulcerated areas for restraint purposes, a 50-ml conical tube can be used as a restraint device that provides access to the rear feet (please ensure holes are drilled in the end of the tube for air exchange).

Fig. 2.32 Performance of hind limb toenail clipping to diminish further self-injury in mice with ulcerative dermatitis. (Reprinted with permission from AALAS. Martel, N, and Careau, C. 2011. *Tech Talk* 16:2–3.)

   - Small scissors, potentially ocular scissors, are recommended to avoid removing more tissue than the toenails (Figure 2.32).
   - Alternatively, nail trims may be performed on rodents during anesthetic recovery from other procedures.
- **Removal of ear tags:** This can be done with wire cutters or using microscissors inserted into the closed tag, followed by opening of the scissor blades to pop open the tags for removal.
- **Sterilization of food and bedding:** To diminish potential hypersensitive elements or contaminants (infectious pathogens).
- **Environmental manipulations:** Could involve the provision of cage enrichments to distract from scratching and a change to hypoallergenic bedding materials.

Scoring systems for UD (Figure 2.33) are helpful, considering that many of these lesions persist for weeks to months without noticeable improvement, despite a variety of treatment attempts (Hampton et al., 2012). Given the variable success rate for eradication of UD, it is recommended that humane endpoints be established if the lesions develop to involve the eyelids, if the lesions are deep enough such that muscle layers are exposed, if greater than 20% of the body is affected,

A	Scratching Number (S number)*	Score
	None	0
	< 5	1
	5–10	2
	> 10	3

* Number of scratches in a 2 minute period.

B	Character of Lesion (COL)	Score
	No lesion present	0
	Excoriations only or one, small punctuate crust (≤ 2 mm)	1
	Multiple, small punctuate crusts or coalescing crust (>2 mm)	2
	Erosion or ulceration	3

C	Length of Lesion**	Score
	0 cm	0
	< 1 cm	1
	1 cm–2 cm	2
	> 2 cm	3

**Length of lesion is determined by measuring the longest diameter of the largest lesion identified. This measurement should involve the lesion only and not cross over clinically normal skin.

D	Regions Affected	Score
	None	0
	Region 2 or 3	1
	Region 2 and 3	2
	Region 1 +/– other affected regions	3

E	Calculated Severity Score
	$[(A + B + C + D) \div 12] \times 100$

Fig. 2.33 Ulcerative dermatitis scoring system. Simplified description for (A) scratching number, (B) character of lesion, (C) length of lesion, and (D) regions affected. Scores from A, B, C, and D are used in the (E) UD scoring system formula to generate the calculated severity score (Reprinted with permission from AALAS. Hampton, AL, Hish, GA, Aslam, MN, Rothman, ED, Bergin, IL, Patterson, KA, Naik, M, Paruchuri, T, Varani, J, and Rush, HG. 2012. *J Am Assoc Lab Anim Sci* 51:586–593.)

if ambulation is affected (particularly through wound contraction around a limb), or the treatments have continued for longer than 4 weeks without improvement (see Chapter 4, "Humane or 'Clinical' Endpoint Considerations").

Vaginal or Uterine Prolapse

- **Cause and impact:** Subsequent to parturition, female mice may exhibit vaginal prolapse, or even uterine prolapse (Figure 2.34), with eversion of tissues outside the external orifice. Typically, although these mice are ill, often exhibiting pyloerection, hunched posture, and respiratory dysfunction, they often are still able to actively nurse their pups. It may be unclear on initial report which tissue has prolapsed; thus, it will be important to involve a veterinary assessment to distinguish a severe rectal prolapse from a vaginal/uterine prolapse, which otherwise may appear to be the same to untrained personnel.
- **Potential treatments:** If the female is in poor health and the prolapse is severe, euthanasia may be warranted, and any pups should be cross fostered to an appropriate dam (see relevant section on this topic).

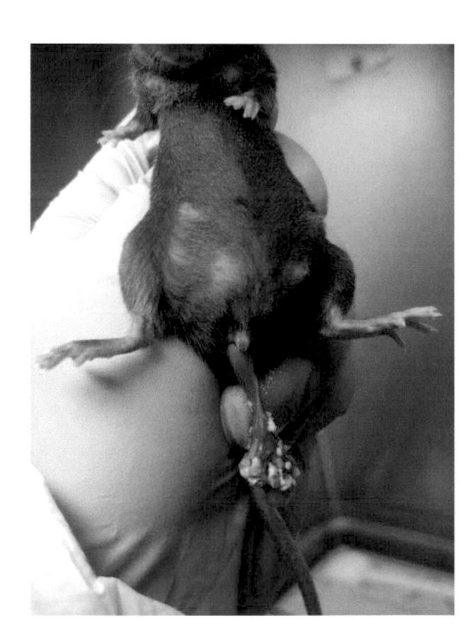

Fig. 2.34 Prolapsed uterus secondary to delivery of a litter; corncob bedding is adhered to the prolapsed tissue. (Image courtesy of University of Pennsylvania, ULAR.)

When the animal is relatively bright, alert, and behaving normally, one can consider correction of the prolapse performed as follows (Castro et al., 2010):

- Reduction of prolapse is done under anesthesia; isoflurane is most preferred for its rapid induction and wide safety margin. Analgesia is provided with carprofen (5.0 mg/kg SC) and fluid replacement (1.0 ml 0.9% NaCL SC).
- Exposed mucosal surfaces should be gently rinsed with saline and cleared of any adhered materials (bedding, etc.) using forceps.
- Topical triple antibiotic ointment should then be applied to the clean tissues, and a lubricated 0.75-inch, 20-gauge polytetrafluoroethylene intravenous catheter (without needle) is placed into the lumen of the uterus and vagina.
- Utilizing the catheter, the exposed tissue is manually manipulated to invert it into the proper anatomic position.
- Purse-string sutures, with 4–0 polyglycolic acid suture, are then placed around the vaginal orifice prior to removal of the catheter to prevent inadvertent recurrence of tissue eversion.
- Postprocedurally, mice can be administered a 3-day course of carprofen (5.0 mg/kg SC) and oral antibiotics.

research-related medical issues

Additional topics concerning laboratory rodent health that warrant further information herein due to their prevalence in contemporary research environments are listed next in alphabetical order to facilitate their location. Under each topic, "background" information is provided, and "potential treatments" offer suggestions about procedures, therapeutic treatments, and further considerations.

Ascites Production

- **Background:** Tumor-producing and ascites-producing cell lines are injected into mice as a method to produce antibodies as a research reagent; this in vivo technique typically is used when no other in vitro method can be implemented. Any animal cell lines should be tested and demonstrated free of murine pathogens and other transmissible agents that potentially could contaminate animal colonies, infect humans, or introduce unwanted experimental variables. If animals are not monitored appropriately, ascites production can be life threatening due to tumor growth, metastatic spread, infiltrative growth, and ultimately respiratory distress.

- **Potential treatments:** The essential treatment for this model is to ensure that mice are monitored for weight changes and development of abdominal swelling due to ascites. Ascites production most commonly occurs within 7 to 14 days after the cells are injected, and the following are suggestions for management of these animals:
  - On the day of and prior to cell inoculation, the mouse should be weighed; this weight is recorded as the "initial weight."
  - Mice that have been inoculated must be weighed at regular intervals; these intervals should be as described in the IACUC protocol and based on the expected rate of abdominal fluid accumulation.
  - Daily clinical observations should be made for assessments related to changes in posture; activity; food and water intake; respiratory patterns (labored, depressed, or accelerated); body condition (e.g., rough hair coat, pale ears or eyes); and severe abdominal distention.
  - Ascites fluid is to be collected before BW becomes 20% greater than the initial weight or abdominal distention leads to significant health problems.
  - Fluid should be harvested following antiseptic preparation of the site using 18- to 22-gauge hypodermic needles. Each time the abdomen is tapped, a fresh disposable needle and syringe are to be used.
  - Warm saline or LRS (2–3 ml SC) may be given at the time the animal is tapped to avoid hypovolemic shock if large volumes of ascitic fluid are removed.
  - Mice should be observed for at least 30 min following a tap.
  - The limit to the number of allowable abdominal taps should be established in the IACUC protocol; the recommended maximum is three taps, with the third tap performed after euthanasia.

- If the abdominal tap does not relieve abdominal distention, the abdomen of the mouse should be gently palpated to determine if distention is due to solid tumor growth.

Clinical signs of hypovolemic shock include hunched posture, roughened hair coat, anorexia, dehydration, weight loss, loss of body condition, inactivity, difficulty in ambulation, pallor of the ears and eyes, tachypnea, and dyspnea. Persistence of these signs after tapping of the abdomen warrants immediate notification of the veterinary staff or consideration of euthanasia of the animal. Mice should be euthanized promptly if ascites fluid becomes blood tinged or turbid or if mice show signs of poor condition, such as huddling, ruffled coat, or inability to reach food and water (IACUC-UPENN, 2010).

Experimental Autoimmune Encephalomyelitis Mouse Models

- **Background:** As a model for multiple sclerosis (MS), the study of experimental allergic encephalitis (EAE) in mice is commonly undertaken. This model may also be referred to as experimental allergic encephalomyelitis, which has the same acronym, EAE. MS is believed to be an autoimmune disease mediated by autoreactive T cells with specificity for myelin antigens. Animals are expected to become weak and may develop an acute, chronic, or relapsing-remitting disease course. In general, the disease progresses with ascending paralysis, dysfunction in normal ability to eat due to lesions in cranial nerves, and lingual paralysis. Animals may lose BW quickly and will likely develop poor overall condition due to inability to masticate food or access provided feedstuffs.
- **Potential treatments:** As disease progresses to excessive weakness, paralysis, and weight loss, softened food can be delivered via oral gavage, and 0.9% NaCl SC can be injected for supportive care. Additional softened food should be placed on the cage floor, water bottles should have elongated sipper tubes to facilitate access by an animal that cannot stand, and urinary bladders should be manually expressed at least twice daily by personnel (Miller and Ito., 2011).

Scoring mechanisms for disease progression may be useful for dctcrmining the humane endpoints of the experiment; more information is available in Chapter 4, "Humane or 'Clinical' Endpoint Considerations" and "Experimental Autoimmune Encephalomyelitis and Demyelinating Disease Model Considerations."

Hemophilic Mouse Models

- **Background:** Mouse models for hemophilia demonstrate prolonged blood clotting times of at least 30 min (the clotting time for a healthy C57BL/6 mouse is less than 5 min). This delay in clotting causes hemorrhage when mice are wounded, potentially due to lesions associated with fighting, experimental injection sites, or spontaneous bleeds. Mice may present with blood in the cage; weakness and lethargy; pallor to the paws, mouth, or gums; and a bloated or swollen appearance, attributed to hemoabdomen.
- **Potential treatments:** Hemorrhage should be prevented by keeping compatible animals in the same cage (with the intent to eliminate aggression and fighting), maintaining pressure over a site that has the potential to bleed for a prolonged time period after a procedure, and handling and monitoring animals carefully to avoid trauma. Bleeding sites may be cleaned and cauterized, with administration of subcutaneous fluids, if necessary (Jones et al., 2005). Silver nitrate sticks are not advisable for use in mice due to adverse skin irritation; studies have shown success in achieving hemostasis in factor VIII knockout mice using ferric sulfate in a gel formulation (supplied in a 1-ml syringe with a small, flat-tipped applicator). This gel formulation has been shown to be superior to the use of fiber cellulose pads, aluminum chloride (both powder and liquid formulations), and ferric subsulfate (both powder and liquid formulations) (Turner et al., 2011).

  In a study of hemophilic PL/J mice (factor IX deficient), D-dimer, a highly specific and reliable test for the diagnosis of thrombotic conditions such as disseminated intravascular coagulation, can be applied for hemophilic mice (Trammell et al., 2006).

Obese Mouse Models

- **Background:** For mice that are genetically engineered or spontaneously develop overweight conditions, considerations should be made for a variety of aspects of their care.

More than 50 different rodent models of obesity are available for use in biomedical research projects (Good, 2005). Some obese mice may grow to 60–70 g BW and therefore, as per housing density standards of the National Research Council (NRC) guide (NRC, 2011), should be housed with fewer cage mates. In particular, for obese rodents, one must take into account the type of housing, the housing conditions (group or singly housed), cage shelf level, the room temperature, and environmental enrichment as all of these issues can have an impact on behaviors, food intake, and BW gain (Good, 2005).

Diabetes, which can accompany the obese phenotype, typically results in increased water consumption and in increased urine production. As well, fluctuations in room temperatures can have an impact on weight gain, fat deposition, and core body temperatures for obese rodents.

Choices of anesthetic agents used for obese rodents should take into account the agent's fat solubility as overweight animals and certain knockout mice will display obesity and reduced sensitivity to certain anesthetic agents (e.g., pentobarbital and tribromoethanol). Performing surgery in these compromised mouse models may result in adverse effects like hypoglycemia, difficulty with dehiscence of incision sites, and anesthesia effects, including delayed recoveries and potentially fatal outcomes due to depressed respiration linked to slower metabolism of drugs.

- **Potential treatments:** Providing supportive care and specialized environments will be best for these obese animals. For animals with evidence of diabetes, more frequent cage changes and provision of more absorbent bedding substrates should be used in the mouse cages to compensate for increased urine production.

  Environmental temperatures and humidity should be closely monitored and potentially lowered for those animals with obesity.

  For those obese mice undergoing surgery, attention to the minimal time for fasting, both presurgical and postsurgical, is key (see Chapter 4, "Fasting Considerations"). To prevent dehiscence of surgical sites, it is recommended to avoid surgical skin clips for this model and instead close incisions with continuous suture patterns, using a minimal suture size. Maintaining the anesthetic dose of isoflurane to no more than 1.0%, with an oxygen flow rate of 0.4 L/min, can facilitate recovery issues and maintain animals at a reasonable depth of anesthesia. As well, supplemental heat should be provided as outlined previously (see relevant sections on this topic). Application of these refinements has been shown to contribute to a survival rate of about 94% for gastrointestinal procedures performed in obese and diabetic mice (Baran and Johnson, 2012).

Opportunistic Infections in Immunodeficient Mouse Models

- **Background:** As a side effect of genetic engineering of mouse strains, many of these animals have mild-to-severe impairment of aspects of the functional immune system. This can lead to overwhelming bacterial infections in these animals, even when housed in barrier-level conditions and when handled by personnel donning appropriate personal protective equipment.

  Klebsiella oxytoca is a common opportunistic agent that has been isolated from urogenital tract infections and abscesses of mice, as well as being identified as the etiology for otitis, keratoconjunctivitis, meningitis, lymphadenitis, and pneumonia (Bleich et al., 2008).

  At times, the source of pathogens may come from experimental agents administered to immunodeficient mice. For example, animals that appear to rapidly lose body condition may be described as "wasting." Wasting disease, with septic arthritis, has been associated with the injection of a *Mycoplasma*-contaminated biological into severely immunodeficient mice (Dodd et al., 2003).

- **Potential treatments:** In the modern facility, it is difficult to maintain a consistently sterile and aseptic housing environment, and it requires a great deal of oversight, labor, and equipment. Autoclaving of all types of equipment and supplies (caging, watering, and food) that come into contact with the immunodeficient mice is of great benefit. In addition, biological agents (to include differing cell types) should be screened, prior to injection, for any opportunists or contaminants.

  Common husbandry practices for immunodeficient transgenic mouse colonies (and for postirradiated animals that have an altered immune system) include the provision of a source of water acidified to less than 3.0 pH as an approach to prevent bacterial contaminations of the water. Laboratory mice provided with acidified water may consume less fluid and have slower growth rates and may weigh less than age-matched controls on untreated water (Craig et al., 1996).

Sterile autoclaved water may provide a reasonable alternative for a fluid source that does not necessarily expose mice to bacteria other than normal flora (Styer et al., 2004). Husbandry management would indicate that the acidified water should be routinely monitored to ensure the pH is not so acidic that it becomes unpalatable to mice. As well, nutritional supplementation in the form of sterile solidified gels may be provided (see Chapter 4, "Nutritional Therapy Considerations").

Radiation Exposure

- **Background:** Xenotransplantation, using immunodeficient mouse models (e.g., NOD/SCID or SCID strains), is a key experimental technique for the study of stem cell biology. The immune system of animals is eliminated using radiation (total body irradiation, TBI) and then the blood cell types may be reconstituted with transplantation. TBI of 3 to 3.5 Gy is used to minimize competition from endogenous bone marrow cells and ensure maximum engraftment of donor hemopoietic cells. In the interim between radiation exposures and acceptance of cell transplants, mice are extraordinarily vulnerable to acquiring infections and developing related disease syndromes.

  Irradiation doses greater than 10 Gy have been reported to cause dental abnormalities in C57BL/6 mice; as well, during a series of xenogeneic transplantation experiments, development of brittle incisor teeth (Figure 2.35) was noted in NOD/SCID mice at approximately 5 to 7 weeks after nonmyeloablative TBI at 3 Gy (Larsen et al., 2006). This abnormality was associated with rapid weight loss in the mice due to the inability to prehend rodent chow.

  Graft-versus-host disease has been described in immunodeficient strains used for study of human tumor biology and adoptive immunotherapy. Animals may present with weight loss, scruffy hair coat, and hunched posture, with a poor prognosis (Figure 2.36).

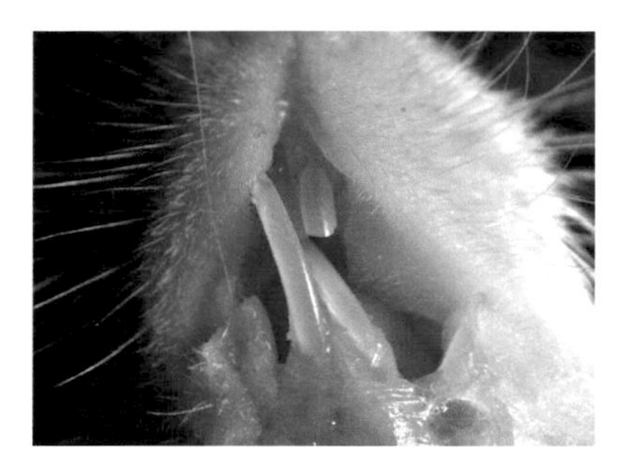

Fig. 2.35 Damage to incisors after nonmyeloablative total body irradiation may complicate NOD/SCID models of hemopoietic stem cell transplantation. Teeth should be trimmed, and animals should have softened food and nutritional support provided on the cage floor. (Reprinted with permission from AALAS. Larsen, SR, Kingham, JA, Hayward, MD, and Rasko, JE. 2006. *Comp Med* 56:209–214.)

Fig. 2.36 Following irradiation, mice may present in poor body condition and be hunched and lethargic. This animal had been gamma-irradiated 3 weeks previously and was provided with daily supportive care, including nutritional supplements, fluids, and antibiotics. (Image courtesy of University of Pennsylvania, ULAR.)

- **Potential treatments:** Multiple aspects of husbandry care (e.g., bioexclusion practices, health monitoring, water quality, use of antibiotics) are to be considered when housing mice that are undergoing irradiation or bone marrow transplantation. Common husbandry practices for postirradiated animals include the provision of a source of water acidified to less than 3.0 pH as an approach to prevent bacterial contamination. It cannot be overemphasized that the first 7 to 10 days after transplantation are the most crucial;

close monitoring of the recipient mice by the laboratory and husbandry staff is highly recommended to identify any possible health problems during this phase and beyond (Duran-Struuck and Dysko, 2009).

Following irradiation, treatments often include administration of oral antibiotics and provision of softened feed (which may be provided with high fat/calories for additional energy). Feed and water consumption can be evaluated to gauge how to use these vehicles for antibiotic delivery.

In one study, consumption of water from sipper tubes for irradiated C57BL/6 mice (dosed at 67 cGy/min) was tracked; it was found that consumption of acidified water dropped by 30% within 1 day of exposure and continued to decrease to about 1.5 ml/day (from 4 ml/day) at 3 weeks. Overall, when wetted chow was available, the intake of fluid directly from sipper tubes was decreased by half. Consumption of acidified water containing ciprofloxacin after irradiation was similar whether grape flavoring or sugar was added for increased palatability (Plett et al., 2008).

Fluid replacement with LRS (1–2 ml SC) immediately following irradiation and twice daily for up to 7 days postirradiation may be instrumental in minimizing loss of animals. LRS can be combined with antibiotics (i.e., enrofloxacin at 85 mg/kg SC twice daily) prior to injection. In contrast, trimethoprim-sulfamethoxazole in the drinking water has *not* been shown to be of overall benefit to mouse health following irradiation (Ramirez et al., 2005).

Early recognition of any subsequent dental abnormalities following TBI is beneficial. As well, trimming of teeth to allow mandibular and maxillary incisor occlusion, coupled with provision of softened chow and supportive care, can improve the dental health of irradiated animals to maintain them through experimental phases.

Humane endpoints should be established and may include cage-side scoring of body postures, appearance, and activity levels. It has been shown in C57BL/6 mice receiving an LD_{50} (median lethal dose) dose of 845 cGy that those animals achieving cage-side scores indicative of declining condition had corresponding mortality rates of 78–100%. The effort should be made to preemptively euthanize animals prior to spontaneous death (Nunamaker et al., 2012).

Following euthanasia of mice with graft-versus-host disease, histologic assessments may reveal underlying dermatitis, colitis and hepatitis, nephritis, arthritis, meningoencephalitis, and vasculitis (Duran-Struuck et al., 2010).

Streptozotocin Induction for Diabetic Models

- **Background:** Streptozotocin (STZ)-induced diabetes in mice is often used to model diabetes mellitus and its complications as well as other pathologies. In studies of diabetes progression and effects of newly developed treatments, experimental results may be difficult to interpret because blood glucose levels (BGLs) of untreated diabetic ICR mice tend to decline substantially during typical experimental time spans of 8 to 11 h. To address this problem, several experimental conditions have been examined that might affect BGL stability, including STZ dose, initial mouse weight, fasting regimens and light–dark cycle within the room. Interestingly, it has been shown that diabetes severity is dependent on initial mouse weight, and that weight loss after diabetes induction is less severe in heavier mice (Dekel et al., 2009).
- **Potential treatments:** BGLs can be stabilized in diabetic mice, particularly for those animals that are not undergoing treatment with insulin, by regulating the amount of food offered to mice during the experiment (Dekel et al., 2009).

Tumor Burden in Mouse Models

- **Background:** Rodent tumor models are extremely prevalent in laboratory animal facilities. Tumor cell suspensions are often implanted subcutaneously over the flank and scapular areas to evaluate growth and immune responses to the transplanted cells; also, tumors can be transplanted following growth in another animal (Workman et al., 2010). Tumors can be induced by chemical carcinogens, by radiation, by surgical anastomoses, and using viral and bacterial agents. Spontaneous tumors can develop in certain strains, and tumor burden should be managed based on how the animal's overall health and body condition fares (Figure 2.37). Note that many rodents may show only subtle signs of clinical disease

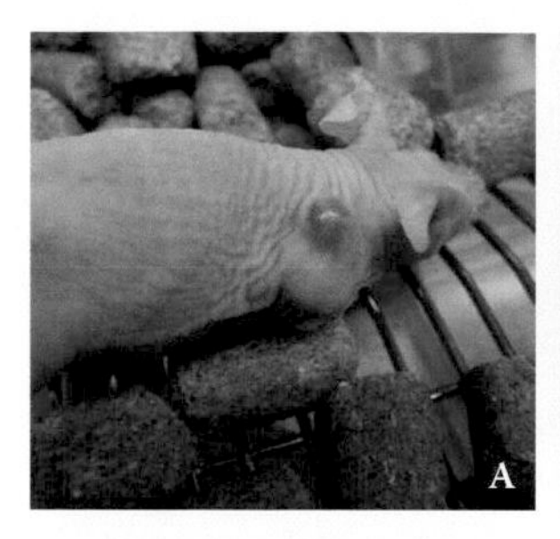
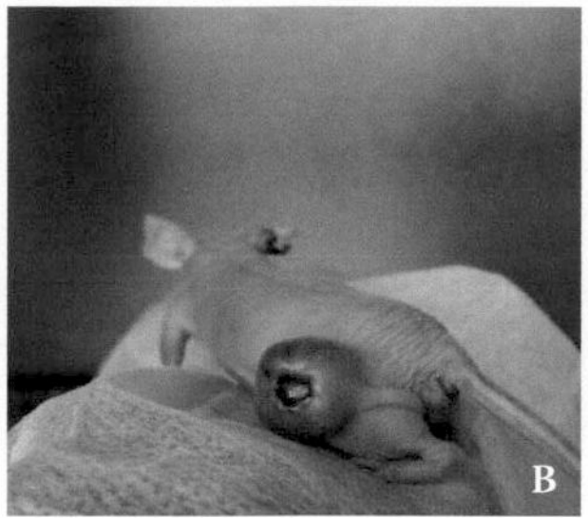
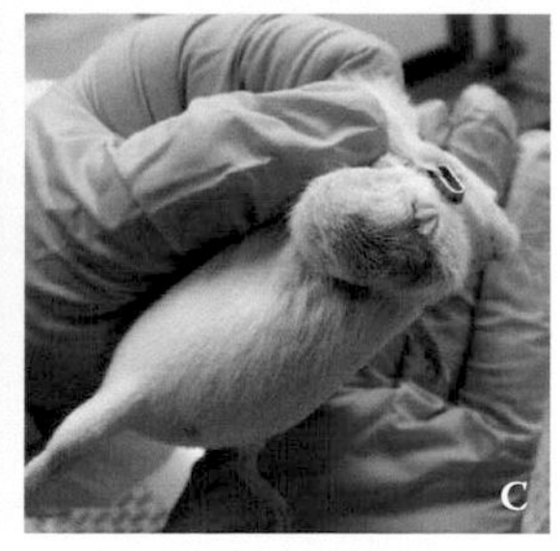

Fig. 2.37 Varying degrees of tumor burden in laboratory mice: (A) focal subcutaneous tumor induced in a nude mouse; (B) large ulcerated tumor following induction in a nude mouse; (C) mass that has encompassed the right forelimb to the extent that ambulation is impaired (this animal would be recommended for euthanasia for humane reasons). (Images courtesy of University of Pennsylvania, ULAR.)

until late in tumor development; often, tumor incidence will develop with increased age and other comorbidities of aging. Comprehensive listings of syngeneic, xenogeneic, and autochthonous tumor models are available for review (Workman et al., 2010).

Ear tag neoplasms have been documented in certain strains of mice, including transgenic animals with a FVB/N background. Related neoplasms have been identified as squamous cell carcinomas (in 9% of mice > 300 days old) (Baron et al., 2005) or fibrosarcomas (Everitt et al., 2002) in locations closely associated with the presence of metal (nickel-copper alloy) ear tags.

Lymphoma may present as a combination of clinical signs, including abdominal or subcutaneous masses, anemia, hunched posture, and poor body condition with ruffled fur. When tumors are not present in subcutaneous locations, scoring systems and schematics for identifying tumor burdens should be established (Paster et al., 2009).

Intracardiac injection of human tumor cells into anesthetized nude mice is an established model of bone metastasis. However, intracardiac injection of some human tumor cell lines causes acute neurologic signs and high mortality, making some potentially relevant tumor cell lines unusable for investigation.

- **Potential treatments:** In genetically modified mice, particular care measures are necessary to ensure detection of unexpected sites of tumor development (Workman et al., 2010). Individual institutional guidance should be followed with respect to size of allowable tumors and increased monitoring of animal health (see Chapter 4, "Tumor Development and Monitoring Considerations"). BW may not vary significantly from baseline as tumor masses develop; therefore, incorporating BCS assessments may be beneficial for monitoring any declines in health related to model development. Other scoring criteria for intra-abdominal tumors could include level of activity after unprovoked and provoked stimulation, fur integrity, posture, breathing effort, abdominal and musculoskeletal palpation, and measurement of circumference of the abdomen and torso (Schenk et al., 2012).

Diagnostics may include biopsy of the mass for histological analysis of cells and tissues, using fine-needle aspirates (FNAs) and impression smears, and potentially ultrasonography for masses in the reproductive tract, abdomen, and mammary glands (Hochleithner and Hochleithner, 2004).

If tumors have ulcerated, meaning that the tumor has outgrown its available blood supply and is now showing tissue necrosis, this should be closely monitored and treated with topical antibiotics and potentially analgesics. Certain studies may warrant the tumor to become ulcerated to test experimental treatments. However, most models that lead to ulcerated tumor formation are typically close to experimental endpoints; therefore, animals should not be kept alive longer than is absolutely necessary to avoid welfare concerns.

Studies have shown that intracardiac injection of tumor cells can induce a hypercoagulable state leading to platelet consumption and thromboemboli formation, and that pretreatment with intravenous injection of low-molecular-weight heparin (LMWH; enoxaparin) blocks this state. In addition, intravenous injection of enoxaparin before intracardiac injection with two different small-cell lung carcinoma lines, H1975 and H2126, dramatically decreased mouse mortality while still generating bone metastases. Therefore, reduction of mortality by pretreatment with LMWH increased the types of cells that can be studied in

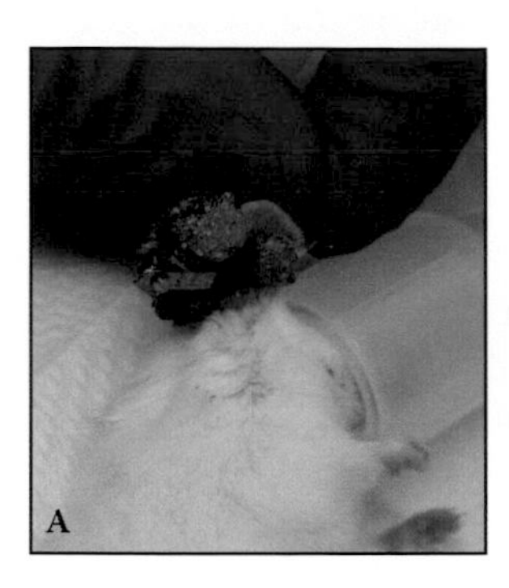

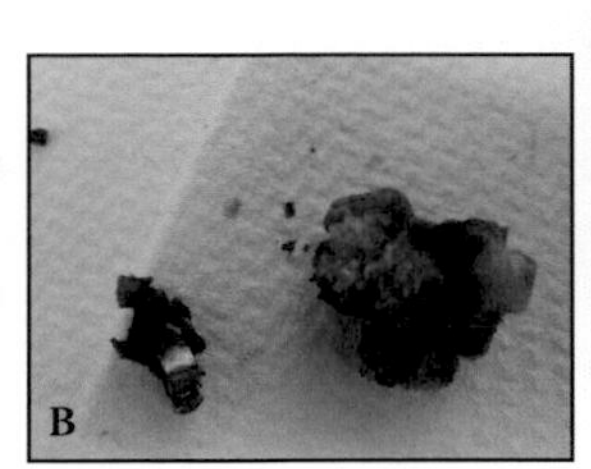

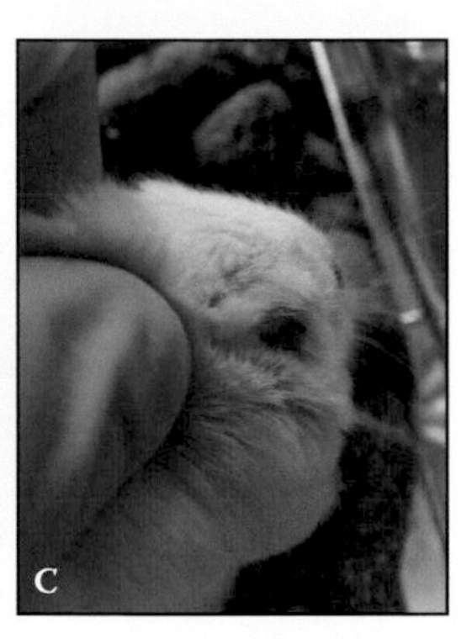

Fig. 2.38 Removal of auricular tumor secondary to ear tag placement (A). Amputation of the pinna, using cordless cautery loop tips, was performed to remove the mass (B). The mouse received triple antibiotic ointment applied topically twice daily for 7 days and meloxicam (5 mg/kg SC) for 3 days postsurgery. The amputation site had healed within 10 days (C). (Images courtesy of University of Pennsylvania, ULAR.)

this metastasis model and decreased the number of animals used (Stocking et al., 2009).

Surgical excision provides a potential cure and is generally more effective than chemotherapy or radiation therapy for those animals with experimentally induced tumors (Figure 2.38); tumor removal results in more cures than all other modalities combined (Mehler and Bennett, 2004).

Mammary neoplasia in mice is almost always malignant; therefore, aggressive resection is recommended if it is not an intended aspect of the experimental model. Unfortunately, recurrence and metastasis of this tumor type is common after surgical resection (Mehler and Bennett, 2004).

For mouse models of melanoma, alternative endpoints have been proposed since the animals may have heavy tumor loads in the absence of other clinical abnormalities (Narver, 2013).

Chemotherapeutics may be administered for treatment, depending on tumor type. Chemotherapy-related fatigue may be present and persistent for up to 4 weeks after treatment, linked to toxicity of the agents administered (Ray et al., 2008). Confirmation of the tumor type should be made by histopathology of tumor tissue samples harvested at surgery or at the humane or experimental endpoint.

Urogenital Disease in Mouse Models

- **Background:** Chronic estrogen exposure has been associated with decreased muscular contraction of the urinary bladder and subsequent urinary retention. This can lead to susceptibility of mice to ascending infections of the urinary tract, potential for bacterial colonization of the bladder (cystitis), and urine scalding around the perineal area.

  Uroliths are typically found more often in male mice, in association with mouse urologic syndrome (MUS). Intact male mice housed on wire flooring have significantly higher rates of MUS than those animals housed directly on bedding; clinical signs include urine staining around the prepuce, edema, perineal ulcerative dermatitis and alopecia, and bladder distention (Everitt et al., 1988).

  Spontaneous struvite urolithiasis has been described in estrogen-treated ovariectomized female nude mice with cystitis induced by *Staphylococcus intermedius*. Concurrent factors of a moist ulcerative dermatitis with associated bacterial infection, as well as predisposition to urine retention and vesicouretal reflux, may result in ascending cystitis and subsequent uroliths. Female nude mice, of normal body condition, have been documented to present with superficial and ulcerative skin lesions around the perineal area and cranial to the tail base.

- **Potential treatments:** Cases of urine scalding in mice may be treated topically with Neosporin ointment once daily or application of Silvadene (similar to treatment for burns). Note that nitrofurazone ointment is a known carcinogen; therefore, it is not recommended for use in rodents (Langlois, 2004). Administration of systemic antibiotics can alleviate cystitis and potentially alleviate urolith burden (Gibbs et al., 2007). A bedding change in mouse cages from corncob substrate to sterilized paper bedding has been linked to a 60% decrease in incidence of severe urogenital disease; this simple husbandry practice change can subsequently lead to an increased survival rate for study animals to reach their experimental endpoints (Simmons et al., 2002).

euthanasia

Euthanasia is the process of inducing painless death in animals. To the greatest extent possible, animals being euthanized should not experience pain, fear, or other significant stress prior to their death. Carbon dioxide (CO_2) exposure or narcosis is a frequently used euthanasia method in the laboratory for small animals due to its rapid onset of action, safety, low cost, and ready availability. Exposure times for carbon dioxide differ dramatically depending on the age of the mouse to be euthanized; mice older than 21 days of age typically require 5 min of exposure (Pritchett et al., 2005). Investigations into the potential advantages of premedicating or anesthetizing mice prior to CO_2 exposure have led to the conclusions that these ancillary approaches do not diminish the behavioral effects of exposure to a low flow rate (defined as displacement of 20% of the cage volume per minute) of CO_2 (Valentine et al., 2012).

Cervical dislocation as a rapid means of physically causing death has been shown to potentially have unacceptably high rates of failure (up to 21%) for mouse euthanasia (Carbone et al., 2012). Injectable and inhalant methods may be preferable to physical means unless individuals have received specific hands-on training. Further discussion is provided in Chapter 4, "Euthanasia Considerations," and in the *AVMA Guidelines for the Euthanasia of Animals* (American Veterinary Medical Association [AVMA], 2013).

references

Abatan, OI, Welch, KB, and Nemzek, JA. 2008. Evaluation of saphenous venipuncture and modified tail-clip blood collection in mice. *J Am Assoc Lab Anim Sci* 47:8–15.

Adamson, TW, Kendall, LV, Goss, S, Grayson, K, Touma, C, Palme, R, Chen, JQ, and Borowsky, AD. 2010. Assessment of carprofen and buprenorphine on recovery of mice after surgical removal of the mammary fat pad. *J Am Assoc Lab Anim Sci* 49:610–616.

Aldrich, J. 2005. Global assessment of the emergency patient. *Vet Clin North Am Small Anim Pract* 35:281–305.

Alworth, LC, and Nagy, T. 2009. Periurethral swelling in a female C3H/HeNCrMTV mouse. *Lab Anim (NY)* 38:147.

American Veterinary Medical Association (AVMA). 2013. AVMA Guidelines for the Euthanasia of Animals: 2013 Edition, pp. 1–102. https://www.avma.org/KB/Policies/Documents/euthanasia.pdf.

Artwohl, JE, Purcell, JE, and Fortman, JD. 2008. The use of cross-foster rederivation to eliminate murine norovirus, *Helicobacter* spp., and murine hepatitis virus from a mouse colony. *J Am Assoc Lab Anim Sci* 47:19–24.

Auestad, N, Korsak, RA, Bergstrom, JD, and Edmond, J. 1989. Milk-substitutes comparable to rat's milk; their preparation, composition and impact on development and metabolism in the artificially reared rat. *Br J Nutr* 61:495–518.

Banks, RE, Sharp, JM, Doss, SD, and Vanderford, DA. 2010. Mice, pp. 73–80. In *Exotic Small Mammal Care and Husbandry*. Wiley-Blackwell, Ames, IA.

Baran, SW, and Johnson, EJ. 2012. Refinement of gastrointestinal procedures in mouse models for obesity and diabetes studies. *J Am Assoc Lab Anim Sci* 51:642.

Barnard, DE, Starost, MF, Teter, B, Yoshizumi, E, Sampugna, J, Morse, B, and Foltz, C. 2006. Dietary effects of the development of ulcerative dermatitis in C57BL/6J Mice. *J Am Assoc Lab Anim Sci* 45:115–116.

Baron, BW, Langan, G, Huo, D, Baron, JM, and Montag, A. 2005. Squamous cell carcinomas of the skin at ear tag sites in aged FVB/N mice. *Comp Med* 55:231–235.

Bazille, PG, Walden, SD, Koniar, BL, and Gunther, R. 2001. Commercial cotton nesting material as a predisposing factor for conjunctivitis in athymic nude mice. *Lab Anim (NY)* 30:40–42.

Beierle, EA, Chen, MK, Hartwich, JE, Iyengar, M, Dai, W, Li, N, Demarco, V, and Neu, J. 2004. Artificial rearing of mouse pups: development of a mouse pup in a cup model. *Pediatr Res* 56:250–255.

Black, CR, Nowland, MH, Hill, VA, and Dombroske, HM. 2011. Neurodegenerative models and the use of gel diet to improve survivability. *J Am Assoc Lab Anim Sci* 50:761–762.

Bleich, A, Kirsch, P, Sahly, H, Fahey, J, Smoczek, A, Hedrich, HJ, and Sundberg, JP. 2008. *Klebsiella oxytoca*: opportunistic infections in laboratory rodents. *Lab Anim* 42:369–375.

Bothe, GW, Bolivar, VJ, Vedder, MJ, and Geistfeld, JG. 2005. Behavioral differences among fourteen inbred mouse strains commonly used as disease models. *Comp Med* 55:326–334.

Burns, C, Gibson, R, and Ehrlich, P. 2005. Methodology for tracheal intubation of rodents during imaging procedures. *Contemp Top Lab Anim Sci* 44:83–84.

Buxbaum, LU, DeRitis, PC, Chu, N, and Conti, PA. 2011. Eliminating murine norovirus by cross-fostering. *J Am Assoc Lab Anim Sci* 50:495–499.

Byrum, R, Alexander, I, Rosa, B, Oberlander, N, Cooper, K, and Rojas, O. 2011. Use of body surface temperature obtained with an infrared thermometer as an early endpoint criterion in orthopoxvirus infection studies. *J Am Assoc Lab Anim Sci* 50:810.

Calderone, L, Grimes, P, and Shalev, M. 1986. Acute reversible cataract induced by xylazine and by ketamine-xylazine anesthesia in rats and mice. *Exp Eye Res* 42:331–337.

Carbone, L, Carbone, ET, Yi, EM, Bauer, DB, Lindstrom, KA, Parker, JM, Austin, JA, Seo, Y, Gandhi, AD, and Wilkerson, JD. 2012. Assessing cervical dislocation as a humane euthanasia method in mice. *J Am Assoc Lab Anim Sci* 51:352–356.

Caro, A, Hankenson, FC, and Marx, JO. 2012. Comparison of thermoregulatory devices during rodent anesthesia and the effects of body temperature on physiologic parameters. *J Am Assoc Lab Anim Sci* 51:685–686.

Carter, DB, Kennett, MJ, and Franklin, CL. 2002. Use of perphenazine to control cannibalism in DBA/1 mice. *Comp Med* 52:452–455.

Castro, PA, Sohn, CS, and Roman, L. 2010. Nonsurgical correction for vaginal/uterine prolapse in mice. *J Am Assoc Lab Anim Sci* 49:690–1.

Ceccarelli, AV, and Rozengurt, N. 2002. Outbreak of hind limb paralysis in young CFW Swiss Webster mice. *Comp Med* 52:171–175.

Chan, MM, and Washington, IM. 2011. Prostaglandin-based treatment of dystocia in the laboratory mouse. *J Am Assoc Lab Anim Sci* 50:804.

Charles, H, Halliday, L, Lang, M, and Fortman, J. 2005. Evaluation of the SnuggleSafe® microwave heatpad in laboratory animal use. *Contemp Top Lab Anim Sci* 44:93–94.

Chew, JL, and Chua, KY. 2003. Collection of mouse urine for bioassays. *Lab Anim (NY)* 32:48–50.

Coman, JL, Buck, WR, Fan, LP, Niquette, AL, and Strasburg, DJ. 2010. Suitability of the submandibular blood sampling technique for serial blood sampling in individual mice. *J Am Assoc Lab Anim Sci* 49:656–657.

Compton, SR. 2008. Prevention of murine norovirus infection in neonatal mice by fostering. *J Am Assoc Lab Anim Sci* 47:25–30.

Cope, MB, Nagy, TR, Fernandez, JR, Geary, N, Casey, DE, and Allison, DB. 2005. Antipsychotic drug-induced weight gain: development of an animal model. *Int J Obes (Lond)* 29:607–614.

Cote, M, Jimenez, A, and Gourdon, J. 2011. Eye problems: why euthanize when it can be treated? *J Am Assoc Lab Anim Sci* 50:753.

Craig, SL, Laber-Laird, KE, Olson, JC, and Swindle, MM. 1996. Effect of water treatment and *Pseudomonas* infection on mortality in irradiated, viral antibody-free mice. *Contemp Top Lab Anim Sci* 35:57–60.

Crowley, ME, Delano, ML, and Kirchain, SM. 2008. Successful treatment of C57Bl/6 ulcerative dermatitis with caladryl lotion. *J Am Assoc Lab Anim Sci* 47:109–110.

Danneman, PJ, Suckow, MA, and Brayton, CF. 2012. *The Laboratory Mouse*, 2nd edition. CRC Press, Boca Raton, FL.

Dardenne, A, Lewis, SM, and La Perle, K. 2011. Unilateral cholesterol granuloma in a male C57BL/6 mouse in a colony with a high incidence of perineal swellings. *J Am Assoc Lab Anim Sci* 50:745.

Dekel, Y, Glucksam, Y, Elron-Gross, I, and Margalit, R. 2009. Insights into modeling streptozotocin-induced diabetes in ICR mice. *Lab Anim (NY)* 38:55–60.

Delgado, R, Gee, LC, Yu, E, and Wallace, J. 2011. Using chlorhexidine to treat dermatitis in mice. *J Am Assoc Lab Anim Sci* 50:755.

Disselhorst, D, Long, L, and Perret-Gentil, MI. 2010. Improving dermatitis and self-injury in rodents through a novel approach to trimming toenails. *J Am Assoc Lab Anim Sci* 49:709.

Dodd, JW, Kelleher, RJ, Menon, M, and Besch-Williford, C. 2003. *Mycoplasma arginini*-associated septic arthritis and wasting in SCID mice. *Contemp Top Lab Anim Sci* 42:66–67.

Doneley, RJ. 2005. Ten things I wish I'd learned at university. *Vet Clin North Am Exot Anim Pract* 8:393–404.

Duarte-Vogel, SM, and Lawson, GW. 2011. Association between hair-induced oronasal inflammation and ulcerative dermatitis in C57BL/6 mice. *Comp Med* 61:13–19.

Dufour, BD, Adeola, O, Cheng, HW, Donkin, SS, Klein, JD, Pajor, EA, and Garner, JP. 2010. Nutritional up-regulation of serotonin paradoxically induces compulsive behavior. *Nutr Neurosci* 13:256–264.

Duran-Struuck, R, and Dysko, RC. 2009. Principles of bone marrow transplantation (BMT): providing optimal veterinary and husbandry care to irradiated mice in BMT studies. *J Am Assoc Lab Anim Sci* 48:11–22.

Duran-Struuck, R, Huang, C, Sachs, DH, Bronson, R, and Spitzer, T. 2010. Graft versus host disease in miniature swine: development of a scoring system to correlate to swine and human clinical disease manifestation. *J Am Assoc Lab Anim Sci* 49:732.

Everitt, JI, Ross, PW, Bobbitt, C, Boorman, GA, Torti, VR, and Butterworth, BE. 2002. Metal ear tag-induced foreign body tumorigenesis in p53+/- mice. *Contemp Top Lab Anim Sci* 41:73.

Everitt, JI, Ross, PW, and Davis, TW. 1988. Urologic syndrome associated with wire caging in AKR mice. *Lab Anim Sci* 38:609–611.

Feldman, SH, McVay, L, and Kessler, MJ. 2006. Resolution of ulcerative dermatitis of mice by treatment with topical 0.2% cyclosporin. *J Am Assoc Lab Anim Sci* 45:92.

Fisher, RS. 1989. Animal models of the epilepsies. *Brain Res Brain Res Rev* 14:245–278.

Forbes, N, Brayton, C, Grindle, S, Shepherd, S, Tyler, B, and Guarnieri, M. 2010. Morbidity and mortality rates associated with serial bleeding from the superficial temporal vein in mice. *Lab Anim (NY)* 39:236–240.

Forbes-McBean, NA, and Brayton, CF. 2012. Mouse urine specific gravity: chemical strip method compared with veterinary refractometer. *J Am Assoc Lab Anim Sci* 51:702.

Fuller, JL, and Sjursen, FH, Jr. 1967. Audiogenic seizures in eleven mouse strains. *J Hered* 58:135–140.

Gibbs, LK, Hickman, DL, Lewis, AD, and Colgin, LM. 2007. *Staphylococcus*-induced urolithiasis in estrogen-treated ovariectomized nude mice. *J Am Assoc Lab Anim Sci* 46:61–65.

Ginty, I, and Hoogstraten-Miller, S. 2008. Perineal swelling in a mouse. Diagnosis: imperforate vagina with secondary mucometra. *Lab Anim (NY)* 37:196–199.

Goelz, MF, Mahler, J, Harry, J, Myers, P, Clark, J, Thigpen, JE, and Forsythe, DB. 1998. Neuropathologic findings associated with seizures in FVB mice. *Lab Anim Sci* 48:34–37.

Goldkuhl, R, Carlsson, HE, Hau, J, and Abelson, KS. 2008. Effect of subcutaneous injection and oral voluntary ingestion of buprenorphine on post-operative serum corticosterone levels in male rats. *Eur Surg Res* 41:272–278.

Good, DJ. 2005. Using obese mouse models in research: special considerations for IACUC members, animal care technicians, and researchers. *Lab Anim (NY)* 34:30–37.

Grindle, S, Garganta, C, Sheehan, S, Gile, J, Lapierre, A, Whitmore, H, Paigen, B, and DiPetrillo, K. 2006. Validation of high-throughput methods for measuring blood urea nitrogen and urinary albumin concentrations in mice. *Comp Med* 56:482–486.

Hamacher, J, Arras, M, Bootz, F, Weiss, M, Schramm, R, and Moehrlen, U. 2008. Microscopic wire guide-based orotracheal mouse intubation: description, evaluation and comparison with transillumination. *Lab Anim* 42:222–230.

Hampton, AL, Hish, GA, Aslam, MN, Rothman, ED, Bergin, IL, Patterson, KA, Naik, M, Paruchuri, T, Varani, J, and Rush, HG. 2012. Progression of ulcerative dermatitis lesions in C57BL/6Crl mice and the development of a scoring system for dermatitis lesions. *J Am Assoc Lab Anim Sci* 51:586–593.

Hankenson, FC, Ruskoski, N, Van Saun, M, Ying, G, Oh, J, and Fraser, NW. 2013. Weight loss and reduced body temperature determine humane endpoints in a mouse model of ocular herpesvirus infection. *J Am Assoc Lab Anim Sci* 52:277–285.

Hawkins, MG, and Graham, JE. 2007. Emergency and critical care of rodents. *Vet Clin North Am Exot Anim Pract* 10:501–531.

Hedrick, CS, Chipps, J, and Hickman-Davis, J. 2009. Development of a standard operating procedure for warming hypothermic mice due to flooded caging. *J Am Assoc Lab Anim Sci* 48:109.

Hickman, DL, and Swan, MP. 2011. Effects of age of pups and removal of existing litter on pup survival during cross-fostering between multiparous outbred mice. *J Am Assoc Lab Anim Sci* 50:641–646.

Hill, LR, Coghlan, LG, and Baze, WB. 2002. Perineal swellings in two strains of mice. *Contemp Top Lab Anim Sci* 41:51–53.

Hochleithner, C, and Hochleithner, M. 2004. Select exotic animal cases using ultrasound. *Exotic DVM* 6.3:53–56.

Hoff, J. 2000. Methods of blood collection in the mouse. *Lab Anim* 29:47–53.

Hong, CC, and Ediger, RD. 1978. Preputial gland abscess in mice. *Lab Anim Sci* 28:153–156.

Horne, D, Saunders, K, and Campbell, M. 2003. Refinement of the saphenous vein blood collection from a mouse without the use of restraining devices or anesthesia. *Contemp Top Lab Anim Sci* 42:121–122.

Hoshiba, J. 2004. Method for hand-feeding mouse pups with nursing bottles. *Contemp Top Lab Anim Sci* 43:50–53.

Hrapkiewicz, K, and Medina, L (eds.). 2007. *Clinical Laboratory Animal Medicine*, 3rd edition. Blackwell, Ames, IA.

Jacobsen, KR, Kalliokoski, O, Hau, J, and Abelson, KSP. 2011. Voluntary ingestion of buprenorphine in mice. *Animal Welfare* 20:591–596.

Jones, K, Tu'akalau, T, Patarroyo-White, S, Liu, T, and Pater, C. 2005. Clinical care and husbandry of hemophilic mouse models. *Contemp Top Lab Anim Sci* 44:74.

Kalliokoski, O, Jacobsen, KR, Hau, J, and Abelson, KS. 2011. Serum concentrations of buprenorphine after oral and parenteral administration in male mice. *Vet J* 187:251–254.

Kastenmayer, RJ, Fain, MA, and Perdue, KA. 2006. A retrospective study of idiopathic ulcerative dermatitis in mice with a C57BL/6 background. *J Am Assoc Lab Anim Sci* 45:8–12.

Klaphake, E. 2006. Common rodent procedures. *Vet Clin North Am Exot Anim Pract* 9:389–413, vii–viii.

Kling, MA. 2011. A review of respiratory system anatomy, physiology, and disease in the mouse, rat, hamster, and gerbil. *Vet Clin North Am Exot Anim Pract* 14:287–337, vi.

Koch, T., and Vincent, V. 2010. Treating rectal prolapse in mice. *J Am Assoc Lab Anim Sci* 49:672.

Kohn, DF, Martin, TE, Foley, PL, Morris, TH, Swindle, MM, Vogler, GA, and Wixson, SK. 2007. Public statement: guidelines for the assessment and management of pain in rodents and rabbits. *J Am Assoc Lab Anim Sci* 46:97–108.

Kurien, BT, Everds, NE, and Scofield, RH. 2004. Experimental animal urine collection: a review. *Lab Anim* 38:333–361.

Kurien, BT, and Scofield, RH. 1999. Mouse urine collection using clear plastic wrap. *Lab Anim* 33:83–86.

Kuster, T, Zumkehr, B, Hermann, C, Theurillat, R, Thormann, W, Gottstein, B, and Hemphill, A. 2012. Voluntary ingestion of antiparasitic drugs emulsified in honey represents an alternative to gavage in mice. *J Am Assoc Lab Anim Sci* 51:219–223.

Langford, DJ, Bailey, AL, Chanda, ML, Clarke, SE, Drummond, TE, Echols, S, Glick, S, Ingrao, J, Klassen-Ross, T, Lacroix-Fralish, ML, Matsumiya, L, Sorge, RE, Sotocinal, SG, Tabaka, JM, Wong, D, van den Maagdenberg, AM, Ferrari, MD, Craig, KD, and Mogil, JS. 2010. Coding of facial expressions of pain in the laboratory mouse. *Nat Methods* 7:447–449.

Langlois, I. 2004. Wound management in rodents. *Vet Clin North Am Exot Anim Pract* 7:141–167.

Larsen, SR, Kingham, JA, Hayward, MD, and Rasko, JE. 2006. Damage to incisors after nonmyeloablative total body irradiation may complicate NOD/SCID models of hemopoietic stem cell transplantation. *Comp Med* 56:209–214.

Lawson, GW. 2010. Etiopathogenesis of mandibulofacial and maxillofacial abscesses in mice. *Comp Med* 60:200–204.

Lawson, GW, Sato, A, Fairbanks, LA, and Lawson, PT. 2005. Vitamin E as a treatment for ulcerative dermatitis in C57BL/6 mice and strains with a C57BL/6 background. *Contemp Top Lab Anim Sci* 44:18–21.

Lumpkins, KC, Swing, S, Emerson, C, Ali, F, and Van Andel, R. 2006. Efficacy of topical chlorhexidine for treatment of ulcerative dermatitis in C57BL/6 mice. *J Am Assoc Lab Anim Sci* 45:94–95.

Luo, Y, Corning, BF, White, WJ, Fisher, TF, and Morin, RR. 2003. An evaluation of a rodent water replacement source. *Contemp Top Lab Anim Sci* 42:119.

Maier, SM, Gross, JK, Hamlin, KL, Maier, JL, Workman, JL, Kim-Howard, XR, Schoeb, TR, and Farris, AD. 2007. Proteinuria of nonautoimmune origin in wild-type FVB/NJ mice. *Comp Med* 57:255–266.

Mangete, ED, West, D, and Blankson, CD. 1992. Hypertonic saline solution for wound dressing. *Lancet* 340:1351.

Martel, N, and Careau, C. 2011. Green clay therapy for mice topical dermatitis. *Tech Talk* 16:2–3.

Mathews, KA, and Binnington, AG. 2002. Wound management using honey. *Compendium* 24:53–60.

Matsumiya, LC, Sorge, RE, Sotocinal, SG, Tabaka, JM, Wieskopf, JS, Zaloum, A, King, OD, and Mogil, JS. 2012. Using the Mouse Grimace Scale to reevaluate the efficacy of postoperative analgesics in laboratory mice. *J Am Assoc Lab Anim Sci* 51:42–49.

McCann, RA, and Mitchel, RE. 1994. A safe and effective method for repeated rectal temperature measurements in mice. *Contemp Top Lab Anim Sci* 33:84–85.

Mehler, SJ, and Bennett, RA. 2004. Surgical oncology of exotic animals. *Vet Clin North Am Exot Anim Pract* 7:783–805, vii–viii.

Miller, AL, and Richardson, CA. 2011. Rodent analgesia. *Vet Clin North Am Exot Anim Pract* 14:81–92.

Miller, S, and Haimovich, B. 2011. Infrared body temperature measurement of mice as an early predictor of death in lipopolysaccharide-induced endotoxic shock. *J Am Assoc Lab Anim Sci* 50:725.

Miller, S, and Ito, K. 2011. Special care for experimental allergic encephalomyelitis mice. *J Am Assoc Lab Anim Sci* 50:819–820.

Mufford, T, and Richardson, L. 2009. Nail trims versus the previous standard of care for treatment of mice with ulcerative dermatitis. *J Am Assoc Lab Anim Sci* 48:78–79.

Murray, SA, Morgan, JL, Kane, C, Sharma, Y, Heffner, CS, Lake, J, and Donahue, LR. 2010. Mouse gestation length is genetically determined. *PLoS One* 5:e12418.

Naff, KA, Van Pelt, C, Craig, S, and Gray, K. 2005. Perianal mass in a female transgenic mouse. *Lab Anim (NY)* 34:31, 32–33.

Narver, H. 2011. Ancillary care improves outcome in a mouse model of spinal muscular atrophy. *J Am Assoc Lab Anim Sci* 50:739–740.

Narver, HL. 2012. Oxytocin in the treatment of dystocia in mice. *J Am Assoc Lab Anim Sci* 51:10–17.

Narver, HL. 2013. Care and monitoring of a mouse model of melanoma. *Lab Anim (NY)* 42:92–98.

National Research Council (NRC). 2011. *Guide for the Care and Use of Laboratory Animals*, 8th edition. National Academies Press, Washington, DC.

Nemzek, JA, Xiao, HY, Minard, AE, Bolgos, GL, and Remick, DG. 2004. Humane endpoints in shock research. *Shock* 21:17–25.

Newsom, DM, Bolgos, GL, Colby, L, and Nemzek, JA. 2004. Comparison of body surface temperature measurement and conventional methods for measuring temperature in the mouse. *Contemp Top Lab Anim Sci* 43:13–18.

Nugent Britt, CC, Blackwell, A, Hall, C, Story, J, Meshaw, A, Corniceli, J, Rosebury-Smith, W, and Hollister, B. 2011. The effects of submandibular vein blood collection on the health of female NCr nu/nu, CD1 and CB.17 SCID mice. *J Am Assoc Lab Anim Sci* 50:791.

Nunamaker, EA, Artwohl, JE, and Fortman, JD. 2012. Endpoint refinement for C57BL/6 mice in total body irradiation studies. *J Am Assoc Lab Anim Sci* 51:646.

Ogeka, S. 2009. A closer look inside rodent water valves. *J Am Assoc Lab Anim Sci* 48:123.

Oglesbee, BL. 2011. Rodents, pp. 544–622. In Oglesbee, BL (ed.), *Blackwell's Five-Minute Veterinary Consult: Small Mammal*, 2nd edition. Wiley-Blackwell, Ames, IA.

Paster, EV, Villines, KA, and Hickman, DL. 2009. Endpoints for mouse abdominal tumor models: refinement of current criteria. *Comp Med* 59:234–241.

Paul-Murphy, J. 1996. Little critters: emergency medicine for small rodents, pp. 714–18, Fifth International Veterinary Emergency and Critical Care Symposium, San Antonio, TX.

Pesapane, R, and Good, DJ. 2009. Seizures in a colony of genetically obese mice. *Lab Anim (NY)* 38:81–83.

Plett, A, Crisler-Roberts, R, Chua, H, Nguyen, N, West, ES, Kohlbacher, KJ, Sampson, CH, Johnson, CA, Juliar, BE, Katz, BM, Brossia, LJ, and Orschell, CM. 2008. Efficacy of ad libitum antibiotics as countermeasures against acute radiation syndrome. *J Am Assoc Lab Anim Sci* 47:171–172.

Pritchett, K, Corrow, D, Stockwell, J, and Smith, A. 2005. Euthanasia of neonatal mice with carbon dioxide. *Comp Med* 55:275–281.

Raabe, BM, Artwohl, JE, Purcell, JE, Lovaglio, J, and Fortman, JD. 2011. Effects of weekly blood collection in C57BL/6 mice. *J Am Assoc Lab Anim Sci* 50:680–685.

Ramirez, HE, Hassan, KM, Brown, GA, Scott, EW, and Sharp, PE, 2005. Fluid therapy benefits whole-body irradiation mice. *Contemp Top Lab Anim Sci* 44:72.

Ray, MA, Trammell, R, Ran, S, and Toth, LA. 2008. Assessment of chemotherapy-related fatigue in mice. *J Am Assoc Lab Anim Sci* 47(6).

Rivera, B, Miller, S, Brown, E, and Price, R. 2005. A novel method for endotracheal intubation of mice and rats used in imaging studies. *Contemp Top Lab Anim Sci* 44:52–55.

Rosenbaum, MD. 2010. Husbandry approaches to alleviate conjunctivitis in neonatal SKH1 hairless mice. *J Am Assoc Lab Anim Sci* 49:699.

Rosenbaum, MD, VandeWoude, S, and Bielefeldt-Ohmann, H. 2007. Sudden onset of mortality within a colony of FVB/n mice. *Lab Anim (NY)* 36:15–16.

Rubino, R, Hanley, G, Winkelmann, C, and Besch-Williford, C. 2004. Perineal hernias in FVB/N mice. *Contemp Top Lab Anim Sci* 43:69.

Schenk, M, Orlando, N, Schenk, N, Haung, CA, and Duran-Struuck, R. 2012. Development of a murine hematopoetic tumor model scoring system. *J Am Assoc Lab Anim Sci* 51:631.

Schowalter, D, Levin, SI, and Zielinski-Mozny, N. 2011. A practical approach to dystocia in mice. *J Am Assoc Lab Anim Sci* 50:747.

Shomer, NH, and Berenblit, H. 2008. Comparison of methods to rewarm hypothermic mice. *J Am Assoc Lab Anim Sci* 47:137–138.

Simmons, JE, Morrow, JE, and Franklin, CL. 2002. Effect of bedding sterilization and type on the incidence of urogenital disease in the estrogenized nude mouse. *Contemp Top Lab Anim Sci* 41:72–73.

Singletary, KB, Kloster, CA, and Baker, DG. 2003. Optimal age at fostering for derivation of *Helicobacter hepaticus*-free mice. *Comp Med* 53:259–264.

Slattum, MM, Stein, S, Singleton, WL, and Decelle, T. 1998. Progressive necrosing dermatitis of the pinna in outbred mice: an institutional survey. *Lab Anim Sci* 48:95–98.

Smith, DE, Blumberg, JB, and Lipman, RD. 1999. Improved survival rates in mice that received prophylactic fluids after carcinogen treatment. *Contemp Top Lab Anim Sci* 38:84–86.

Smith, PC, Zeiss, CJ, Martin-Escalante, D, Herrick, CA, and Bottomly, K. 2003. Pruritic dermatitis associated with *Demodex musculi* in transgenic mice. *Contemp Top Lab Anim Sci* 42:111.

Spoelstra, EN, Ince, C, Koeman, A, Emons, VM, Brouwer, LA, van Luyn, MJ, Westerink, BH, and Remie, R. 2007. A novel and simple method for endotracheal intubation of mice. *Lab Anim* 41:128–135.

St. Claire, MB, Sowers, AL, Davis, JA, and Rhodes, LL. 1999. Urinary bladder catheterization of female mice and rats. *Contemp Top Lab Anim Sci* 38:78–79.

Stocking, KL, Jones, JC, Everds, NE, Buetow, BS, Roudier, MP, and Miller, RE. 2009. Use of low-molecular-weight heparin to decrease mortality in mice after intracardiac injection of tumor cells. *Comp Med* 59:37–45.

Strom, JO, Theodorsson, A, Ingberg, E, Isaksson, IM, and Theodorsson, E. 2012. Ovariectomy and 17beta-estradiol replacement in rats and mice: a visual demonstration. *J Vis Exp* e4013.

Styer, CM, Griffey, SM, and Kendall, LV. 2004. Normal flora contamination of water in mice receiving acidified and autoclaved water. *Contemp Top Lab Anim Sci* 43:51.

Suckow, MA, Danneman, P, and Brayton, C. 2001. *The Laboratory Mouse*. CRC Press, Boca Raton, FL.

Swan, M, McCrea-Gant, E, and Hickman, D. 2010. Conjunctivitis with blepharitis and enophthalmos in a colony of hairless mice housed in individually ventilated caging. *J Am Assoc Lab Anim Sci* 49:717–718.

Taylor, DK. 2007. Study of two devices used to maintain normothermia in rats and mice during general anesthesia. *J Am Assoc Lab Anim Sci* 46:37–41.

Taylor, DK, Lorch, G, Silva, K, Miller, J, Nicholson, A, Vonder Haar, R, Sperling, L, King, LE, and Sundberg, JP. 2005. Study of the etiology of spontaneous alopecia and ulcerative dermatitis in C57BL/6J laboratory mice. *Contemp Top Lab Anim Sci* 44:86.

Taylor, DK, Rogers, MM, and Hankenson, FC. 2006. Lanolin as a treatment option for ringtail in transgenic rats. *J Am Assoc Lab Anim Sci* 45:83–87.

Taylor, R, Hayes, KE, and Toth, LA. 2000. Evaluation of an anesthetic regimen for retroorbital blood collection from mice. *Contemp Top Lab Anim Sci* 39:14–17.

Tomlinson, R, Abramson, M, Edwards, K, Reed, G, Diehl, A, and Medrano, J. 2004. Comparing the effectiveness of mandibular versus retro-orbital blood collection. *Contemp Top Lab Anim Sci* 23:53.

Tonsfeldt, E, Hickman, DL, and Van Winkle, DM. 2007. An accessible and humane approach to mouse intubation. *J Am Assoc Lab Anim Sci* 46:102–103.

Toth, LA, and Gardiner, TW. 2000. Food and water restriction protocols: physiological and behavioral considerations. *Contemp Top Lab Anim Sci* 39:9–17.

Trammell, RA, Brooks, M, Cox, L, Ding, M, Wagenknecht, DR, Rehg, JE, McIntyre, JA, and Toth, LA. 2006. Fatal hemorrhagic diathesis associated with mild factor IX deficiency in pl/J mice. *Comp Med* 56:426–434.

Truett, GE, Walker, JA, and Baker, DG. 2000. Eradication of infection with *Helicobacter* spp. by use of neonatal transfer. *Comp Med* 50:444–451.

Turner, PA, McNally, TJ, Johnson, LA, and Cunliffe-Beamer, TL. 2011. Evaluation of several hemostatic agents to control bleedign in factor VIII knockout mice. *J Am Assoc Lab Anim Sci* 50:780.

Ullman-Cullere, MH, and Foltz, CJ. 1999. Body condition scoring: a rapid and accurate method for assessing health status in mice. *Lab Anim Sci* 49:319–323.

Valentine, H, Williams, WO, and Maurer, KJ. 2012. Sedation or inhalant anesthesia before euthanasia with CO_2 does not reduce behavioral or physiologic signs of pain and stress in mice. *J Am Assoc Lab Anim Sci* 51:50–57.

Van Loo, P, Kruitwagen, C, Van Zutphen, L, Koolhaas, JM, and Baumans, V. 2000. Modulation of aggression in male mice: influence of cage cleaning regime and scent marks. *Animal Welfare* 9:281–295.

Van Loo, PL, Van Zutphen, LF, and Baumans, V. 2003. Male management: coping with aggression problems in male laboratory mice. *Lab Anim* 37:300–313.

Vernau, KM, and LeCouteur, R.A. 2009. Seizures and status epilepticus, pp. 414–419. In Silverstein, D, and Hopper, K (eds.), *Small Animal Critical Care Medicine*. Saunders, St. Louis, MO.

Wagner, D, Curtin, L, Baker, S, Passini, M, Savage, ST, and Lee-Parritz, D. 2011. Specialized husbandry improves welfare in a mouse model of spinal muscular atrophy. *J Am Assoc Lab Anim Sci* 50:764–765.

Ward, GM, Cole, K, Faerber, J, and Hankenson, FC. 2009. Humidity and cage and bedding temperatures in unoccupied static mouse caging after steam sterilization. *J Am Assoc Lab Anim Sci* 48:774–779.

Wiedmeyer, CE, Ruben, D, and Franklin, C. 2007. Complete blood count, clinical chemistry, and serology profile by using a single tube of whole blood from mice. *J Am Assoc Lab Anim Sci* 46:59–64.

Williams-Fritze, MJ, Carlson Scholz, JA, Zeiss, C, Deng, Y, Wilson, SR, Franklin, R, and Smith, PC. 2011. Maropitant citrate for treatment of ulcerative dermatitis in mice with a C57BL/6 background. *J Am Assoc Lab Anim Sci* 50:221–226.

Workman, P, Aboagye, EO, Balkwill, F, Balmain, A, Bruder, G, Chaplin, DJ, Double, JA, Everitt, J, Farningham, DA, Glennie, MJ, Kelland, LR, Robinson, V, Stratford, IJ, Tozer, GM, Watson, S, Wedge, SR, and Eccles, SA. 2010. Guidelines for the welfare and use of animals in cancer research. *Br J Cancer* 102:1555–1577.

Yajima, M, Kanno, T, and Yajima, T. 2006. A chemically derived milk substitute that is compatible with mouse milk for artificial rearing of mouse pups. *Exp Anim* 55:391–397.

Yardeni, T, Eckhaus, M, Morris, HD, Huizing, M, and Hoogstraten-Miller, S. 2011. Retro-orbital injections in mice. *Lab Anim (NY)* 40:155–160.

Zuurbier, CJ, Emons, VM, and Ince, C. 2002. Hemodynamics of anesthetized ventilated mouse models: aspects of anesthetics, fluid support, and strain. *Am J Physiol Heart Circ Physiol* 282:H2099–2105.

critical care management for laboratory rats

introduction

The laboratory rat continues to be broadly studied as a model species for investigating disease pathophysiology. Rats are second only to laboratory mice in the number used for biomedical research; fortunately, due to their similarities, many treatment applications described for mouse models can be extrapolated for administration to rats. Three main advantages to using laboratory rats for experimental purposes for studies are their comparatively larger size, coupled with their 3-week gestation period and production of large numbers of offspring. In addition, there are known scientific areas in which the laboratory rat is more similar to humans than the mouse, including the vascular system, the complexity of the rat brain for study of cerebral disorders, and the enzymatic ability of the rat liver to metabolize drugs. The continuing increase in rat genetics data and the rat genome have led to centralization of this information; these resources are highlighted further in Chapter 5. General information on working with laboratory rats is best reviewed in the companion text *The Laboratory Rat* (Sharp and Villano, 2012). Further background information on strains, stocks, and genotypes can also be obtained by visiting the originating vendor source websites, and additional resources are highlighted in Chapter 5.

overall assessments

When assessing laboratory rats, as described for laboratory mice, it will be essential to compile a thorough database of information on health status, research project enrollment, and any potential procedures or treatments already administered. Additional routine aspects of any critical care "history" (see details in Chapter 1) should include the background strain, gender, and age to gain the greatest portfolio of information prior to finalizing differential diagnoses. Further, any changes to the animal's environmental and housing parameters should be reviewed for contribution to the clinical signs. These can include macroenvironmental influences of lighting, noise and vibration, and temperature and humidity of the room; as well, the microenvironment of the cage (diet, water source, housed singly or with other rats, bedding substrate) is to be considered with respect to maintenance of animal health.

general medical approaches to physical examination and health assessments

If rats present in a critically poor state, as determined by the listing of abnormal health conditions in Chapter 1 (see Table 1.2), it will be essential to minimize stress and prioritize clinical interventions into smaller diagnostic and treatment steps. A medical record information template and sick animal reporting sheet are provided and can be used for any rats noted to be in less-than-optimal health (see Chapter 1, Figures 1.1 and 1.2). Typical values for biologic parameters in rats are presented in Table 3.1. The size of the average adult laboratory rat can range widely depending on gender, with females typically lighter weight than males (overall ranging from 250 to greater than 500 g). Despite the increased body size when compared to laboratory mice, there still are strong limitations on the ability to precisely quantify temperature, heart rate, and respiration rate without the use of telemetry or other special equipment in rats.

Physical Examination

Knowledge of the appearance of a laboratory rat in clinically normal health will be key to ensure recognition of abnormal clinical signs should they appear. Visual examination of the animal is the first

Table 3.1: Miscellaneous Parameters for the Laboratory Rat

Parameter	Value
Lifespan	2.5–4.0 years
Puberty	50 ± 10 days
Gestation	21–23 days
Male body weight	450–520 g[a]
Female body weight	250–400 g[a]
Blood volume	57–70 ml/kg = 17.1–21 ml total/300-g rat
Food intake	5–6 g/100 g BW/day
Water intake (ml/100 g BW/day)	10–12 ml/100 g BW/day
Packed cell volume (PCV)	35–57%
Glucose	80–300 (mg/dl)[b]
Body temperature (rectal)	35.9–37.5°C (96.9–99.5°F)
Respiratory rate	70–150 breaths per minute
Heart rate	250–600+ beats per minute

Source: Adapted from Banks, RE, Sharp, JM, Doss, SD, and Vanderford, DA. 2010. Rats, pp. 81–92. In *Exotic Small Mammal Care and Husbandry.* Wiley-Blackwell, Ames, IA; and Sharp, PE, and LaRegina, MC. 1998. *The Laboratory Rat.* CRC Press, Boca Raton, FL.

[a] Weights will vary depending on diet, age, stock, or strain.

[b] Enzyme values are dependent on collection method and may be influenced by anesthesia.

step in assessing the overall physical condition of the laboratory rat and should be done with the rat in its home cage or housing setup prior to manual examination. Rats in critically poor health may benefit from access to supplemental heat and increased oxygen (flow rate 1–2 L/min) exposure (Klaphake, 2006).

As with standard handling practices for laboratory animals, to prevent transmission of potential human pathogens and unwanted exposure to animal allergens, *fresh disposable gloves should be donned prior to manual restraint and handling of laboratory rats.* Handling of rats throughout their time in the research environment will assist with their acclimation to this interaction with personnel. Data on a handling approach called "tickling" suggest that stress associated with handling and intraperitoneal (IP) injections is minimized using this technique (Cloutier et al., 2010). Playful handling includes gentle manipulation and petting of the rats for a few minutes, both before and after a procedure, for the patient to develop further acclimation to its human handlers.

Rats are difficult to lift by the scruff and will often vocalize and struggle against this type of restraint. Instead, a firm grip on the tail base of the rat will facilitate lifting the animal out of the cage (Figure 3.1), followed by placement of the animal on a stable surface

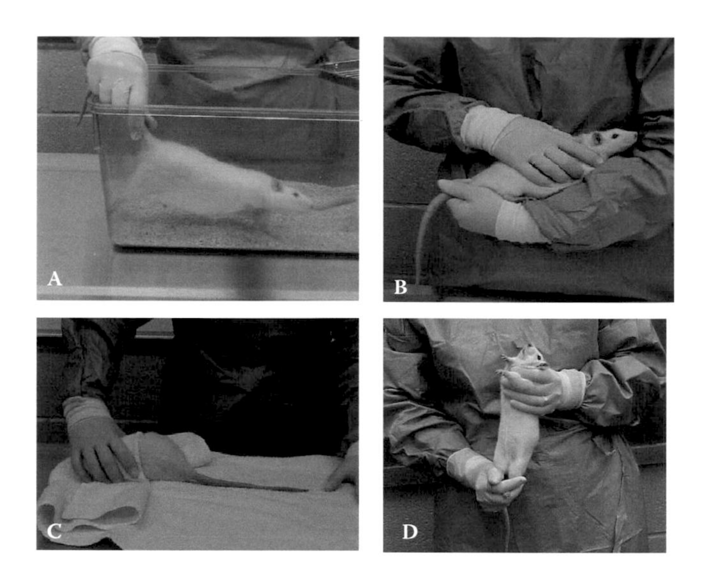

Fig. 3.1 Retrieval of rat from cage can be conducted using a firm grip at the tail base (A) to lift the animal and then placing it into the crook of the staff member's arm (B). Alternatively, the rat is placed on a firm surface or benchtop and can be calmed by covering the eyes with a towel, which also serves to protect the handler from animal bites (C). The two-handed method of restraint allows for a partner to administer a physical examination or treat with therapeutics (D). (Images courtesy of University of Michigan, ULAM.)

(e.g., on a laboratory bench, flow hood surface, or in the crook of the handler's arm held adjacent to the torso). Gentle two-handed restraint is preferred to best assess overall condition.

Handling allows for the ability to closely observe skin and hair coat conditions, any ocular discharge or abnormalities, tooth overgrowth, abnormal masses, or unusual presentations in the anogenital region. Physiological aspects, like *body weight* (BW), *activity*, and *behavior assessments*, are useful to measure and monitor serially. *Heart and lung sounds* should be auscultated and can be performed using a pediatric stethoscope. *Hair coat quality* should be reviewed regarding location of areas of alopecia (baldness), open or closed wounds, or poor grooming. In addition, *respiratory status* (difficult or labored breathing with a more frequent/diminished rate than expected) should be evaluated. Relative *perfusion status*, ascertained by the color of mucous membranes (and potentially by color check of ear and tail tissue), reflects the transport of fluid and oxygen in blood to meet metabolic needs. Collectively, these physiologic measures provide a crude assessment of the "ABCs" (airway/breathing/circulation) of critical care medicine.

Rats in a critical state may need to be sedated to perform physical assessments while minimizing stress responses. Gentle palpation of the abdomen, using a pincer technique with the thumb and forefinger, should help to confirm pregnancy in females and further identify abnormalities like abdominal masses and growths, mammary tissue enlargement, or bladder distention. Finally, the particular experimental use of the rat, as described and approved in the approved proposal to the IACUC (Institutional Animal Care and Use Committee), must be considered, and any adverse effects of the experimental procedures should be documented.

Body Condition Scoring

Assessing general body condition, as described in mice, remains an excellent semiquantitative tool to apply toward rats for assessing health status. The use of a body condition score (BCS) scale (generally on a range from 1, for wasted or emaciated, to 5, for obese) is greatly enhanced by the incorporation of available cartoon diagrams that represent each score on the scale (Hickman and Swan, 2010). The uniformity of the diagrams can be exceptionally valuable for assessments done by a laboratory animal group with variable levels of experience in working with rats (Figure 3.2).

Overall percentages of weight loss should be tracked in rats yet may or may not indicate a loss of health condition, depending on the disease model and whether the animals are expected to develop spontaneous or experimental tumors or other syndromes. Typically, weight loss of more than 20–25% from preexperimental baseline may warrant critical care measures and potentially euthanasia, depending on institutional policies.

Clinical Assessments of Ill Health and Pain in Rats

Rats, though predators of some animals, are considered as prey species in the biomedical research environment. As such, similar to mice, they are conditioned to suppress overt painful behaviors, particularly when being handled. The following are clinical assessments of ill health and pain in rats (Kirsch et al., 2002, Kohn et al., 2007, Miller and Richardson, 2011, Roughan and Flecknell, 2004):

- Vocalization, particularly when handled or a painful area is palpated

BC 1
Rat is emaciated

- Segmentation of vertebral column prominent if not visible.
- Little or no flesh cover over dorsal pelvis. Pins prominent if not visible.
- Segmentation of caudal vertebrae prominent.

BC 2
Rat is under conditioned

- Segmentation of vertebral column prominent.
- Thin flesh cover over dorsal pelvis, little subcutaneous fat. Pins easily palpable.
- Thin flesh cover over caudal vertebrae, segmentation palpable with slight pressure.

BC 3
Rat is well-conditioned

- Segmentation of vertebral column easily palpable.
- Moderate subcutaneous fat store over pelvis. Pins easily palpable with slight pressure.
- Moderate fat store around tail base, caudal vertebrae may be palpable but not segmented.

BC 4
Rat is overconditioned

- Segmentation of vertebral column palpable with slight pressure.
- Thick subcutaneous fat store over dorsal pelvis. Pins of pelvis palpable with firm pressure.
- Thick fat store over tail base, caudal vertebrae not palpable.

BC 5
Rat is obese

- Segmentation of vertebral column palpable with firm pressure; may be a continuous column.
- Thick subcutaneous fat store over dorsal pelvis. Pins of pelvis not palpable with firm pressure.
- Thick fat store over tail base, caudal vertebrae not palpable.

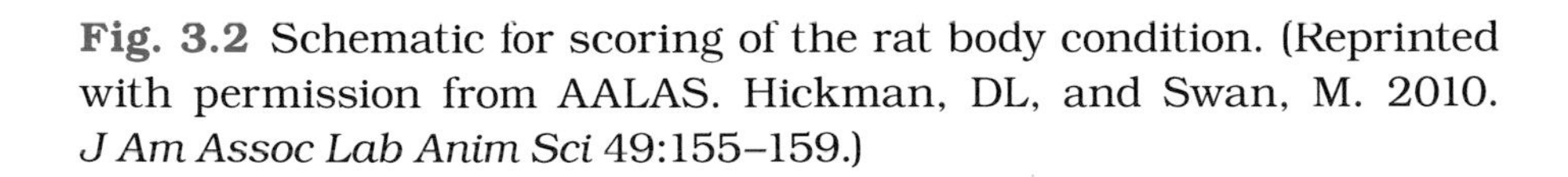

Fig. 3.2 Schematic for scoring of the rat body condition. (Reprinted with permission from AALAS. Hickman, DL, and Swan, M. 2010. *J Am Assoc Lab Anim Sci* 49:155–159.)

- Reduced grooming or piloerection, leading to a "ruffled fur" appearance
- Reduced level of spontaneous and exploratory (sniffing, rearing) activity to the point that rats may not be moving ("moribund condition")
- Hunched posture with "guarding" of abdomen and reduced mobility
- Squint-eyed appearance (either unilateral or bilateral)
- Increased aggressiveness on handling; may bite
- Porphyrin secretions (located around the eyes and nose); distinguish from bloody discharge by use of black light exposure to the secretion type (porphyrin will "glow"; blood will not)
- Distanced from cage mates
- Reduced body condition, likely secondary to reduced nutritional intake or experimental model resulting in muscle wasting and weight loss
- Self-mutilation (excessive licking, biting, scratching) of the painful area
- Abdominal writhing, increased back arching, falling or staggering, poor gait, and twitching
- Palpation of unexpected masses

Monitoring Frequency

Similar to the laboratory mouse, a detailed and descriptive plan for scheduled monitoring of rats both before and after any planned experimental procedures, including the provision of therapeutic treatments and supportive care, should be included in the IACUC protocol submission. Investigators should be aware that as the potential for pain/distress in research animals rises, there should be an increasing intensity of monitoring and frequency of observations.

Objective Scoring Systems

Professional and clinical judgments are essential for the evaluation of an animal's well-being and are critical to the ultimate decision of euthanasia for humane reasons. As well, objective data-based approaches to predicting imminent death, when developed for specific experimental models, should facilitate the implementation of timely euthanasia before the onset of clinically overt signs of moribund state (Toth and Gardiner, 2000). As described for mice, scoring systems are one way in which rats can be monitored throughout an experiment, and systems can be developed for individual experimental needs.

Novel approaches to pain assessment in laboratory rats have been described based on coding of facial expressions, referred to as the Rat Grimace Scale (RGS) (Langford et al., 2010, Sotocinal et al., 2011). Rats are the most common animal model for preclinical pain research (Mogil, 2009), and the RGS was used to improve quantification of pain in three common algesiometric assays: intraplantar instillation of complete Freund's adjuvant, intra-articular kaolin/carrageenan administration, and laparotomy. In contrast to the grimace scale in mice, control rats display distinct bulging of the nose and cheek regions; with pain, the bridge of the nose flattens and elongates, further causing the whisker pads to flatten. This action unit of "nose/cheek flattening" shows the highest correlation with the presence of pain in the rat. The other action units, measured on the 0–2 scale, include orbital tightening and ear and whisker changes (Figure 3.3).

Overall, quantifying pain by facial changes provides a practical clinical assessment in that it can be performed in real time by trained investigators, animal technicians, and veterinary staff (Sotocinal et al., 2011).

veterinary care measures

Administration of Fluids

Dehydration is often present in rats that have pain or are unwell and may be assessed by performing a skin tent or gentle pinch of scruff over the scapulae of the rat and assessing the time that passes for skin to return to normal placement. A prolonged return time indicates a degree of dehydration that should be ameliorated. Fluid administration through the subcutaneous (SC) route should be the least-invasive way to provide supplemental fluid support to the sick rat. Injections into the peritoneal cavity have been evaluated to determine how best to avoid accidental puncture of the cecum, and it has been shown that by *avoiding* the left lower side of the abdomen and injecting into the right lower side, the cecum is not affected (Coria-Avila et al., 2007). Intravenous (IV) injections

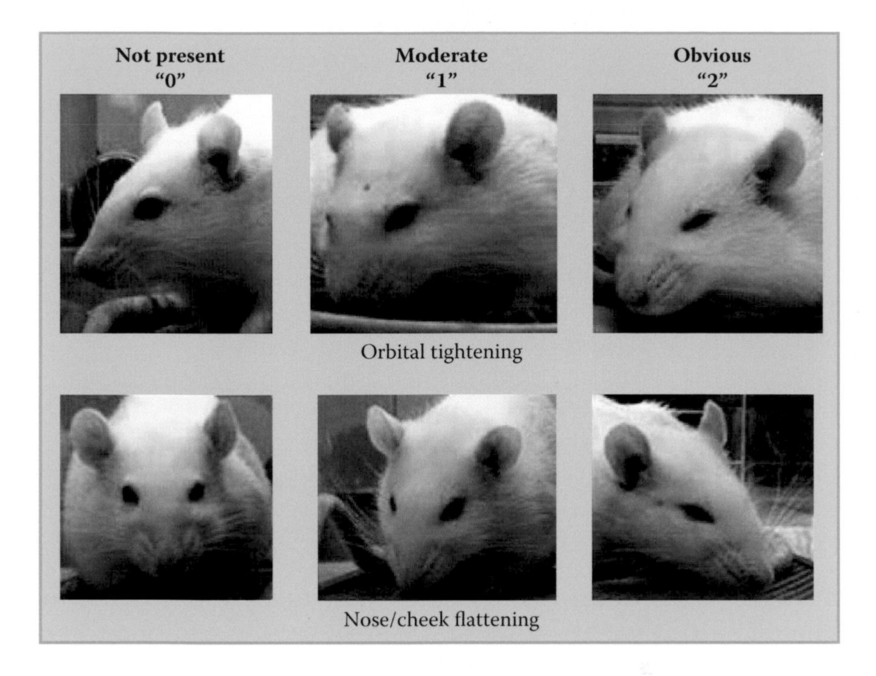

Fig. 3.3 Representative photographs of certain action units of the Rat Grimace Scale for a rat at baseline (facial grimacing not present, 0; a rat with moderate facial grimacing, 1; and a rat with obvious facial grimacing, 2) (Reprinted with permission from Biomed central open access. Sotocinal, SG, Sorge, RE, Zaloum, A, Tuttle, AH, Martin, LJ, Wieskopf, JS, Mapplebeck, JC, Wei, P, Zhan, S, Zhang, S, McDougall, JJ, King, OD, and Mogil, JS. 2011. *Mol Pain* 7:55.)

can be performed for rats using the femoral, jugular, or tail vein, with animals appropriately sedated for access to the larger vessels (Figure 3.4); incisions may be required to gain access to the vessel of choice (Turner, Brabb, et al., 2011). Attempts at refinements for smaller-volume dosing have identified the superficial penile vein of the rat as an option for intravenous injections (Shapiro et al., 2010). As a reminder, the beneficial effect of playful handling (tickling) for rats is strongest when provided both immediately before and after injection (Cloutier et al., 2010).

Water and fluid replacement sources are gaining in popularity, expanding from products initially developed as sustainable fluid sources for the duration of rodent shipping and transport. The provision of these water replacements, in disposable single-use containers, is typically done on the cage floor for rapid access by those animals in ill health. These supplementary fluid sources, when combined

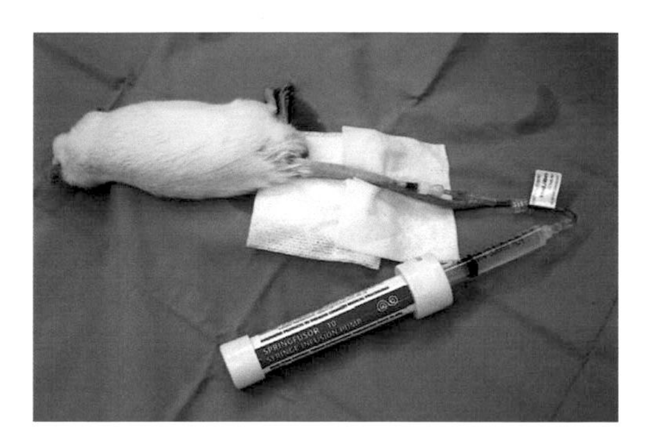

Fig. 3.4 Fluids may be administered intravenously to anesthetized rats using the lateral tail vein and a syringe infusion pump. (Reprinted with permission from AALAS. Turner, PV, Pekow, C, Vasbinder, MA, and Brabb, T. 2011. *J Am Assoc Lab Anim Sci* 50:614–627.)

with food, can maintain the health of rodents for several days in the absence of routine water sources (Luo et al., 2003). Additional critical care considerations for nutritional support, fluid administration, and available products are provided in Chapter 4.

Blood Sampling

Blood sampling, or venipuncture, choices in rats may be influenced by sampling site, anesthetic agent, and method of collection (Fitzner Toft et al., 2006). Sampling allows for testing of serum chemistry parameters, as well as complete blood counts. Suggested sampling sites and further commentary are provided in Table 3.2.

As a guide, the volume of blood taken during a single survival collection should be limited to that needed, not in excess of 10% total blood volume (TBV) in rats; this also may be defined as a limit of about 1.0 ml/100 g BW (Sharp and LaRegina, 1998). For example, for 1% of BW to be withdrawn, 2.5 ml could be sampled at a single time point from a 250-g rat. Following sampling of 1% BW volume, replacement fluid therapy (0.5–1.0 ml SC or IP of sterile isotonic fluid) should be provided.

Retro-orbital blood sampling may be performed with animals under anesthesia but has been associated with subsequent lens opacities and a higher outcome of clotted samples, as compared to other methods (Mathieu, 2011). Other alternative sampling sites in rats include the lateral saphenous vein (Figure 3.5), the sublingual vein, and tail vessels.

Table 3.2: Recommended Sampling Sites and Related Information for Blood Collection in Rats

Anatomical Site	Anesthesia?	Approximate Range of Volume Collected	Comments
Lateral saphenous vein	Not required	Up to 1% of BW	
Sublingual vein	Not required	50–100 μl	
Lateral tail vein	Not required	Up to 1% of BW	
Tail clip	Recommended	Up to 1% of BW	<2 mm of distal end of tail should be clipped; analgesia should be considered
Jugular vein	Recommended	Up to 1% of BW	
Submandibular	Not required	Up to 1% of BW	
Retro-orbital vasculature	Required	Up to 1% of BW	
Cardiac	Required	3+ ml	Terminal procedure only

Source: Modified from University of Pennsylvania, ULAR.

The sublingual vessel can be accessed with the animal unanesthetized and securely restrained, similar to a basic hold used for performing an oral gavage. The mouth will open wide enough to expose the sublingual vasculature, and the oral cavity should be rinsed gently with saline or water and dried prior to sampling. Using a 25- or 23-gauge needle, the vein is punctured, and blood is collected via drip method into the appropriate collection tube. Gauze can then be packed under the tongue to achieve hemostasis (Kohlert, 2012). Care must be taken with any method to ensure that structures surrounding the sampling site are not injured. As well, digital pressure should be applied to achieve hemostasis following blood collection. For critically ill animals, the tail sectioning method may be used by making a transverse perpendicular incision at the tip of the tail (Liu et al., 1996).

The submandibular technique initially performed in mice has been adapted for use in rats. Rats should be lifted by the scruff tightly behind the ears to include as much loose skin as possible; alternatively, sedation will assist with the ability to limit mobility for this procedure. Once lifted by the scruff, the insertion point for the lancet should be located on the jawline, directly below the lateral canthus of the eye. Lancet sizes vary, but a 5.5-mm lancet has been used successfully in rats for this collection method. As with all sampling approaches, hemostasis must be achieved following sampling, typically through manual pressure over the lancet site of insertion (Arzadon, 2011).

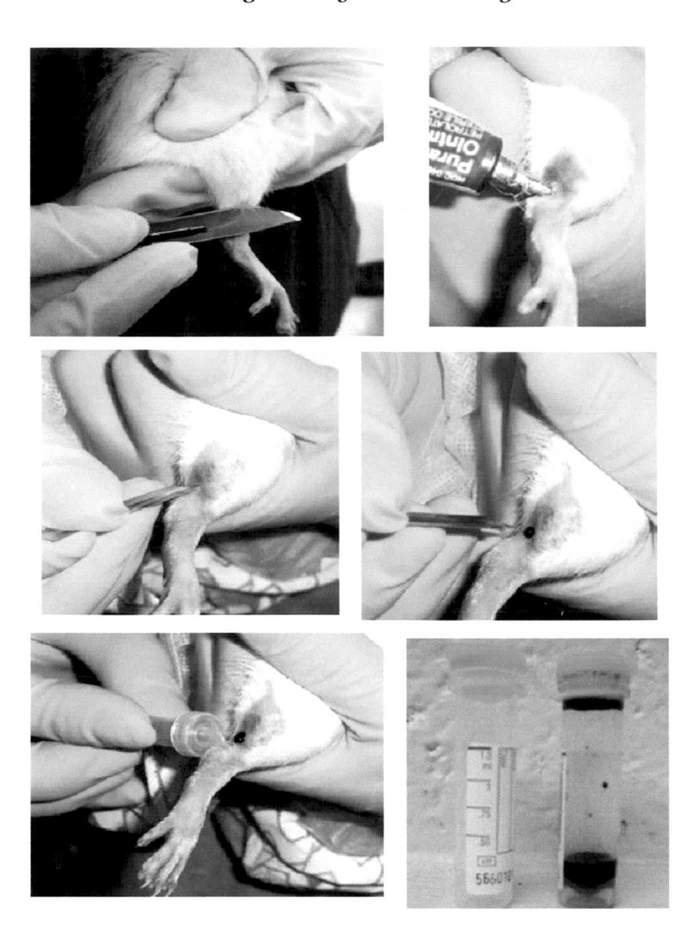

Fig. 3.5 Lateral saphenous vein sampling in the rat. The exterior leg is shaved delicately with a scalpel blade, and lubricant is applied over the vessel to allow blood to bead for collection (top); the vessel is pierced with a medical lancet and blood allowed to pool over the vessel (middle); the hematocrit tube is used to directly capture drops of blood for later submission for serum chemistry and complete blood counts (bottom).

Jugular venipuncture may also be utilized, and it has been performed successfully in conscious and sedated animals using one-handed restraint of the rat. If sedating the animal, the thorax should be shaved to access the jugular vein and then swabbed with an alcohol pad. The collection needle should be inserted above the nipple line at a 20° angle using a minute vacuum, until blood is drawn into the collection tube. It is recommended to avoid continuous pressure on the syringe plunger so the vessel will not collapse. Gauze should be held in place over the blood draw site until hemostasis is achieved (Zeleski et al., 2011).

Body Temperature Monitoring

Often, simple handling of ill rats will provide some indication of whether they are excessively cool or warm to the touch, but body temperature variations have to be extreme for manual detection. Rectal thermometer probes require gentle placement and positioning during procedures and may be more readily utilized in sedated rats, given their larger body size compared to mice. Microchip transponders that provide identification as well as thermometry are also useful for rodents (Bio Medic Data Systems, Seaford, DE).

Body temperature monitoring is critical for animals that are scheduled to undergo prolonged anesthesia; the goal is to mitigate hypothermia associated with experimental and surgical procedures. Suggestions to ameliorate hypothermia would include incubators and warm water bags, as well as Mylar-backed drapes, to reduce radiant heat loss. In addition, warm water recirculation or forced-air (Bair Hugger®, Arizant Healthcare, Eden Prairie, MN) blankets may be beneficial and synergistically effective when coupled with Mylar-backed draping (Koch et al., 2008).

No matter the type of draping used, personnel should ensure that draping allows for viewing of animals to ensure appropriate anesthetic administration and respiratory monitoring. Adverse incidents involving unrecognized surgical fires occurring below the level of a typical blue surgical drape have been described for the rat (Caro et al., 2011). Drapes coupled with a heat-emitting gel pad (Figure 3.6) can provide acceptable thermal support in the rat (Taylor, 2007).

Bone Marrow Access

Bone marrow aspiration from rats has been described as a method to obtain antemortem cell samples. A minimally invasive approach harvesting marrow from the femur (Figure 3.7) spares the knee joint and serves to minimize potential damage to the musculature of the quadriceps (Ordodi et al., 2006).

Endotracheal Intubation

Endotracheal intubation (Figure 3.8) can be readily performed in the rat using either a method of blind access or a strong external light

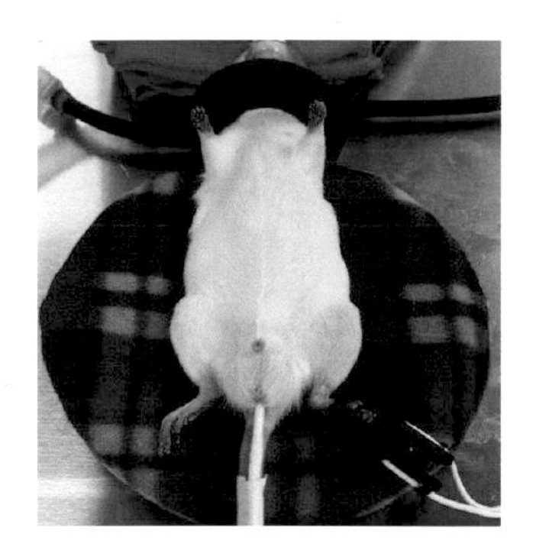

Fig. 3.6 Reusable heating pad (SnuggleSafe® Microwave Heatpad, West Sussex, UK) used for thermal support of anesthetized rats, with manufacturer's cover intact. (Reprinted with permission from AALAS. Taylor, DK. 2007. *J Am Assoc Lab Anim Sci* 46:37–41.)

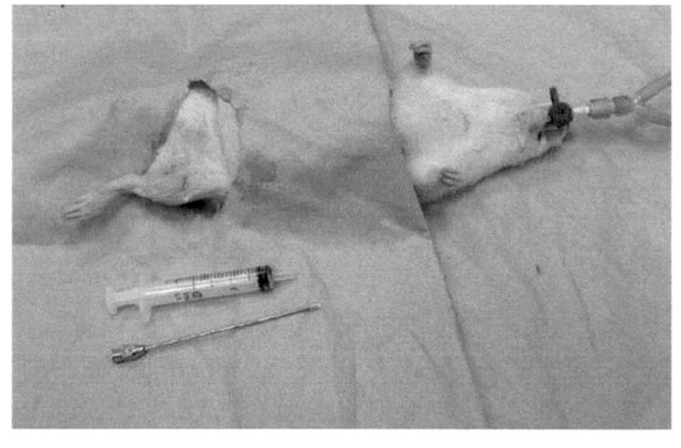

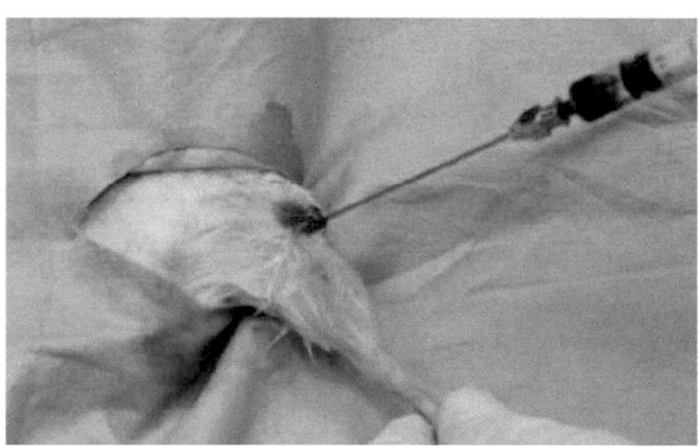

Fig. 3.7 Bone marrow harvesting in an anesthetized rat. In preparation for bone marrow harvesting, the rat should be anesthetized and intubated; then, the thigh area should be shaved and disinfected for the procedure (top). A 14-gauge needle and a 2-ml syringe are the required instruments for harvest; the needle pierces the anterior face of the thigh above the knee joint and is advanced into the femur prior to aspiration of cell sample (bottom). (Reprinted by permission from Macmillan Publishers Limited. Ordodi, VL, Mic, FA, Mic, AA, Tanasie, G, Ionac, M, Sandesc, D, and Paunescu, V. 2006. *Lab Anim (NY)* 35:41–44.)

source that penetrates the skin to illuminate the larynx and facilitate intubation. Other options for endotracheal tubing can be fashioned from standard 2-ml syringes and a light source to illuminate the oropharyngeal cavity, providing easy localization of the larynx (Molthen, 2006, Ordodi et al., 2005). If the rat is in respiratory distress, intubation should be undertaken with caution; however, it can be attempted in animals weighing more than 100 g (Paul-Murphy, 1996). Note that in the critically ill rat, intubation may be challenging and should be

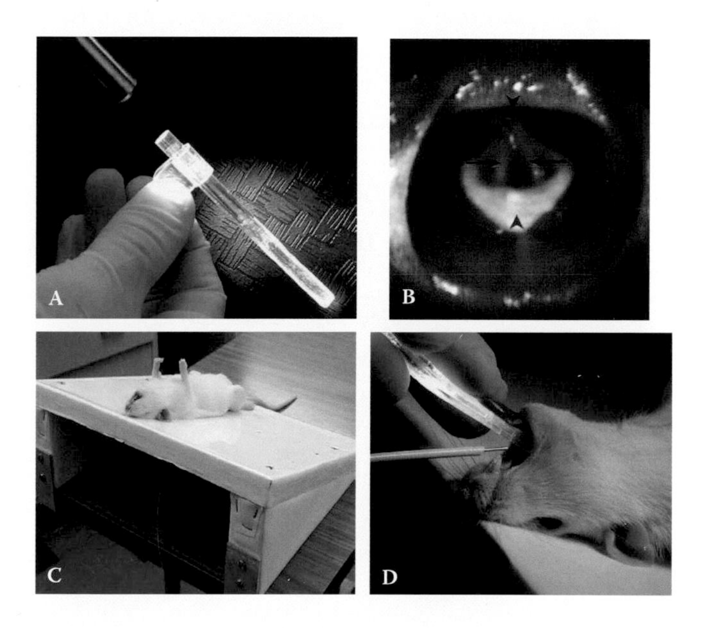

Fig. 3.8 Intubation in the rat. (A) This photograph of a laryngoscope with a light source incident on the proximal end illustrates how light is transmitted to the distal surfaces. (B) Image of laryngeal opening of a rat showing the epiglottis (red arrowhead), arytenoid cartilages (black arrows), and caudal margin of the soft palate (black arrowhead). Visual appearance is similar in the mouse. (C) An anesthetized rat positioned and restrained on inclined plane. (D) The laryngoscope is positioned in the oral cavity to provide visualization of the larynx. The tongue is grasped against the shaft of the laryngoscope. The stylet and tracheal tube are shown before being inserted into the oral cavity. Note the relative position of stylet within the tracheal tube. (Reprinted with permission from AALAS. Molthen, RC. 2006. *J Am Assoc Lab Anim Sci* 45:88–93; and Rivera, B, Miller, S, Brown, E, and Price, R. 2005. *Contemp Top Lab Anim Sci* 44:52–55.)

attempted only as a last resort to gain airway access if tracheostomy cannot be performed (see relevant section in Chapter 4).

Injections and Oral Administration

Injections can be performed routinely using multiple routes (Table 3.3) for the rat, including subcutaneous (SC), intradermal (ID), intraperitoneal (IP), intratracheal (IT), and intravenous (IV), as described in Chapter 1. Intravascular access ports, typically placed surgically in the subcutaneous space over the shoulder area for laboratory rats, can be accessed using Huber needles for substance administration (Figure 3.9). The intramuscular route (IM) can also be more readily used with small-volume injections, as compared to the mouse.

Table 3.3: Recommendations for Injection Dose Limits Based on Weight of Laboratory Rats

	Injection Limits (ml/kg)					
Route	**PO**	**SC**	**IP**	**IM**	**IV (Bolus[a])**	**IV (Slow)**
Dose (ml/kg)	10	5	10	0.1	5	20
Weight (kg)						
0.200	2.0 mL	1.0 mL	2.0 mL	0.02 mL	1.0 mL	4.0 mL
0.225	2.2	1.1	2.2	0.02	1.12	4.5
0.250	2.5	1.2	2.5	0.02	1.25	5.0
0.275	2.7	1.3	2.7	0.02	1.35	5.5
0.300	3.0	1.5	3.0	0.03	1.5	6.0
0.325	3.2	1.6	3.2	0.03	1.6	6.5
0.350	3.5	1.7	3.5	0.03	1.75	7.0
0.375	3.7	1.8	3.7	0.03	1.85	7.5
0.400	4.0	2.0	4.0	0.04	2.0	8.0
0.425	4.2	2.1	4.2	0.04	2.1	8.5
0.450	4.5	2.2	4.5	0.04	2.25	9.0
0.475	4.7	2.3	4.7	0.04	2.35	9.5
0.500	5.0	2.5	5.0	0.05	2.50	10.0
0.525	5.2	2.6	5.2	0.05	2.6	10.5
0.550	5.5	2.7	5.5	0.05	2.75	11.0
0.575	5.7	2.8	5.7	0.05	2.85	11.5
0.600	6.0	3.0	6.0	0.06	3.0	12.0
0.625	6.2	3.1	6.2	0.06	3.1	12.5
0.650	6.5	3.2	6.5	0.06	3.25	13.0
0.675	6.7	3.3	6.7	0.06	3.35	13.5
0.700	7.0	3.5	7.0	0.07	3.50	14.0

Source: Modified from University of Pennsylvania, ULAR.

[a] A bolus is a larger dose given over a shorter period of time.

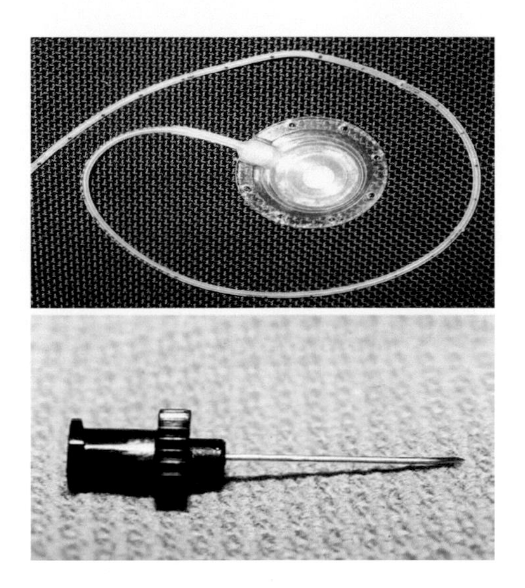

Fig. 3.9 Access device for administration of substances in rats. Vascular access port (top); noncoring Huber needle (bottom). (Reprinted with permission from AALAS. Turner, PV, Pekow, C, Vasbinder, MA, and Brabb, T. 2011. *J Am Assoc Lab Anim Sci* 50:614–627.)

Administration of drugs and fluids can also be done by oral gavage; however, animals must be healthy enough to drink from syringes. This method of fluid/drug administration is less invasive and is easily undertaken once rats are trained to the syringe (Figure 3.10) (Atcha et al., 2010, Turner, Brabb, et al., 2011). If repeated oral (PO) dosing is required, acclimation of rats to handling and to gavaging (with up to 5 ml/kg of a control aqueous material) will help to diminish chronic stress from the procedures (Turner et al., 2012).

Creative approaches to disguising medications in palatable substances, like analgesics in gelatin desserts (Jell-O®), have been described for rats (Flecknell, Orr, et al., 1999, Flecknell, Roughan, et al., 1999). Attention should be paid to the differences in dosing that may be required if delivering oral medications compared to subcutaneous administrations; as well, rats may need to acclimate to the novel substance prior to drug incorporation (Martin et al., 2001).

Urine Sampling

Clinically healthy rats will often dribble urine, which allows for a free-catch sample (Klaphake, 2006). Slight pressure can be applied

Fig. 3.10 Oral dosing of laboratory rats. Rats readily drinking galantamine (0.5 mg/kg) by the novel syringe-dosing method after an acclimation training period (left); animal voluntarily consuming nutritional supplement from a syringe (right) (Reprinted with permission from AALAS. Atcha, Z, Rourke, C, Neo, AH, Goh, CW, Lim, JS, Aw, CC, Browne, ER, and Pemberton, DJ. 2010. *J Am Assoc Lab Anim Sci* 49:335–343; and Turner, PV, Brabb, T, Pekow, C, and Vasbinder, MA. 2011. *J Am Assoc Lab Anim Sci* 50:600–613.)

over the bladder to assist with expression of urine, and one should ensure that an appropriate sterile receptacle is positioned to collect the sample (Kurien et al., 2004). Critically ill rats should be stabilized prior to attempting urinary catheterization if urine collection by other methods has been unsuccessful. Urinary catheterization should only be performed on anesthetized animals. Aseptic technique (see Chapter 4, "Perioperative Care Considerations") and an atraumatic approach should be used during placement of a urinary catheter. Prior to insertion of the catheter, the external urinary orifice should be gently cleansed using a disinfecting (e.g., chlorhexidine) solution. The individual performing the catheterization is advised to don sterile surgical gloves, use a sterile catheter, and apply a small amount of sterile water-soluble lubricant on the external urinary orifice. Additional sterile lubricant should be applied in a thin layer to cover the surface of the urinary catheter for ease of insertion into the urinary orifice, as described for mice (St. Claire et al., 1999). The diameter of the urinary catheter should be the minimum that can be inserted into the bladder and still prevent urinary leakage around the catheter.

The anatomy of the female rat is unique in that the urinary orifice is external and just anterior to the vaginal opening. Adult female rats can be catheterized with number 50 polyethylene (PE) tubing

(2.9 French), a 3.5-French TomCat catheter, or a number 4 Coude urethral catheter that has a bend to the tip of the catheter. This bend facilitates passage of the catheter through the urethra. A guidewire can be threaded through the PE tubing to increase the rigidity of the catheter. Care should be taken that the tip of the guidewire does not extend past the end of the catheter. Guidewires can be made of stainless steel surgical wire and coated with a water-soluble lubricant to ease placement and removal from the PE tubing. The approximate distance from the external urinary orifice to the neck of the bladder for 200-g female rats is 17 mm (St. Claire et al., 1999).

If urine concentration tests are to be performed in rats, personnel should be aware that dehydration may be secondary to any prolonged water deprivation necessary for this type of assay. Studies that assessed clinical condition and BW, at a frequency of every 2 h beginning 16 h after food and water deprivation were initiated, showed a mean BW loss of 8% at 16 h and nearly 10% at 22 h. Clinical dehydration was noted by 22 h, whereas appropriate urine concentration was noted at 16 h. Therefore, it is recommended to complete the rat urine concentration test within a 16-h period to maintain welfare of the animals for this procedure (Kulick et al., 2005).

abnormal, critical, and emergent conditions

Categories of laboratory rodent health concerns are discussed in alphabetical order to facilitate location by the reader. Under each topic, "cause and impact" has been provided, and "potential treatments" offer suggestions about procedures, therapeutic treatments, or husbandry and environmental alterations. Every attempt has been made to provide citations from the literature for evidence-based medical outcomes.

It is essential to note that certain abnormal conditions can be assessed and treated in laboratory rats by similar methods to those done for laboratory mice; see relevant sections in Chapter 2 for the following:

- **Abdominal swelling**
- **Abscessation**
- **Cage flooding**
- **Cannibalization**
- **Cross fostering**

- **Dystocia**
- **Fight wounds**
- **Fractures/orthopedic problems**
- **Hemorrhage**
- **Mortality (sudden death)**
- **Ocular lesions**
- **Respiratory distress**
- **Trauma**
- **Ulcerative dermatitis**

For those health concerns that list drug therapy options, please refer to the rodent formulary provided in Appendix C for additional details on dosages and route of delivery.

Burns

- **Cause and impact:** Burns may be the unfortunate outcome of improper surgical preparation of the animal with alcohol-based disinfectants. Care should be taken to ensure that animal skin that is prepped with alcohol is not then inadvertently ignited by cautery tools during surgery. Smoke inhalation and superficial and partial-thickness burns have been documented to result from this sort of accident (Figure 3.11), the severity of which may be missed due to obstructive surgical draping (Caro et al., 2011). Burns may result in blistering and skin necrosis, shock, and secondary bacterial infections.
- **Potential treatments:** Animals should be provided with oxygen supplementation and stabilized following smoke inhalation. The extent of burn damage should be assessed and any wounds cleansed, debrided, and covered with a topical antibacterial cream, like silver sulfadiazine. To prevent secondary bacterial infection, treatment with systemic antibiotics should be considered. As well, warmed subcutaneous fluids should be provided to offset shock and prevent dehydration. Pain management should be a priority, with opioids or nonsteroidal anti-inflammatory drugs (NSAIDs) provided throughout the duration of the initial healing phases. The prognosis will depend on the extent and thickness of the burned area; aggressive management and monitoring are advised, and euthanasia may be warranted.

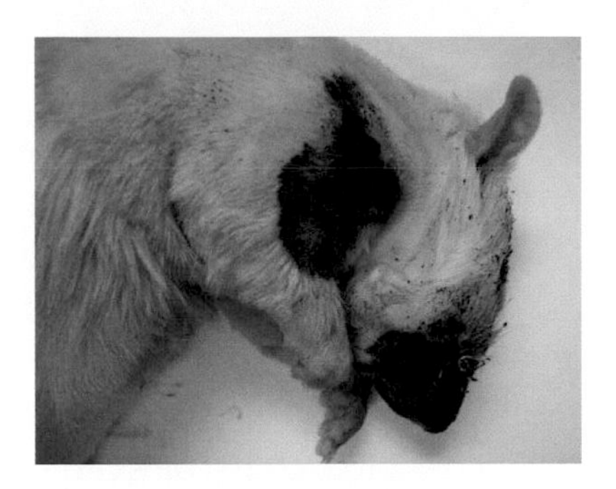

Fig. 3.11 Surgical burn in a rat undergoing a procedure to emulate rotator cuff injury. The animal was covered completely by the surgical drape, and the combination of alcohol applied to the surgical site and cautery resulted in a surgical fire that singed the whiskers, muzzle, and incision margins in the rat. Due to the degree of smoke inhalation, this rat was ultimately euthanized. (Images courtesy of University of Pennsylvania, ULAR.)

Catheter Infections

- **Cause and impact:** Rats, more commonly than mice, are catheterized using vascular access ports (Figure 3.9), typically into the jugular vein, for chronic administration of any variety of test substances. Any indwelling catheter has the potential to serve as a nidus of infection, leading to systemic illness with clinical signs of decreased body condition and activity and altered behavior. As well, localized inflammation can occur at the skin surface and in the subcutaneous space with development of pustular material and a threat to catheter patency, particularly during chronic studies.
- **Potential treatments:** Frequent attention to catheter care is the key prevention strategy against infections. Catheters should be flushed at least twice weekly, once at the time of treatment and again 3 days later. Skin overlying the vascular access port can be cleansed with a chlorhexidine scrub, alternated with dilute povidone–iodine solution. Gentle manual restraint of the rats will permit access to the port site; a noncoring Huber needle can then be inserted through the skin and into the port reservoir. It is recommended to flush the catheter with a volume of about 0.2 ml saline, followed by 0.2 ml heparinized dextrose. Following treatment administration, an additional 0.2 ml saline can

be injected to purge the catheter line. Catheters should be "locked" with an anticoagulant to assist with patency; this material can include heparinized dextrose or may involve 500 IU of heparinized glycerol (Wachtman et al., 2006, Weiner et al., 2012).

Anecdotally, flushing the catheter with saline every 3 to 7 days has been evaluated and found to have negative impacts on the ability to withdraw blood samples (Luo et al., 2004).

Should rats present with signs of systemic disease, antibiotics can be administered, along with supplemental fluids and potential removal of the catheter from the animal. If the rat is intended only for a study requiring chronic catheterization, it may be best to euthanize the septic rat in lieu of aggressive attempts at treatments.

Malocclusion (Incisors) and Caries

- **Cause and impact:** Incisor overgrowth may occur secondary to congenital tooth patterns or may relate to lack of appropriate caging materials for gnawing, softened foods, or a genetic predisposition. Rats that have difficulty prehending food will be anorectic, lose weight, and typically show a decreased BCS within a relatively rapid (~24- to 48-h) time frame.

  Dental caries may develop spontaneously in certain transgenic rat strains; animals should be monitored for signs of anorexia and potential pain secondary to development of caries (Nishijima et al., 2007).
- **Potential treatments:** Treatment includes trimming overgrown teeth to a normal length and alignment. A diamond blade, Dremel®, and dental burr are recommended tools for use on rat dentition. Care should be taken not to crack or split the teeth, which could potentially cause pain and lead to tooth root infections. For valuable rats with potential oral pain secondary to teeth abnormalities and caries, anesthetic extractions of affected teeth may be a potential treatment. Subsequent provision of softened nutritional supplements and wetted chow may be necessary to maintain body condition (see Chapter 4, "Nutritional Therapy Considerations").

  Special attention should be given to the potential for malocclusion in aged rats (especially noted in Wistar rats) during long-term rodent studies as the increased incidence may be detrimental to maintenance of health and general well-being (Dontas et al., 2010).

Moribund/Weak/Paralyzed

- **Cause and impact:** Hind limb weakness (paresis) and paralysis in laboratory rats are associated with trauma, weakness, and dysfunction of the musculoskeletal and nervous systems, neurologic disease models, adverse surgical outcomes, or trauma that may be due to environmental or experimental influence. Neoplasia and nonneoplastic diseases, such as osteoarthritis, bone fractures, or peripheral neuropathies, may also occur, particularly in aged rats (Ceccarelli and Rozengurt, 2002).

  Rat models of spinal cord injury are commonly implemented for biomedical research; in addition to the induced spinal cord lesions, the injured rats may experience alterations of the liver, lung, bladder, and kidneys (Robinson et al., 2012).

  Be aware that rats may also self-injure (autophagia or autotomy) as a consequence of spinal cord or peripheral nerve injury research, associated with altered mobility and pain (Figure 3.12).
- **Potential treatments:** Rats found in a weakened and potentially unresponsive state should be provided with ancillary and supportive care of warmed subcutaneous fluids (2–4 ml 0.9% NaCl) and softened bedding substrates, nutritional supplementation (including softened food pellets on the cage bottom), and supplemental heat, until the level of responsiveness is determined. It is critical to increase the frequency of monitoring and determine humane endpoints that eliminate prolonged suffering for paralyzed or moribund rats (see Chapter 4, "Humane or 'Clinical' Endpoint Considerations").

  For those rat models of spinal cord injuries, researchers should be aware that suprapubic bladder catheterizations performed postinjury will not prevent development of renal abnormalities in rats; therefore, manual expression of the bladder should be performed two to three times daily to eliminate urine accumulations (Robinson et al., 2012).

  Increased observations and monitoring should be done for those animals with self-inflicted lesions. If autophagia has

Fig. 3.12 Self-injury in an adult male Sprague-Dawley rat. Following a left sciatic nerve transection and 48 h of postoperative analgesia, this rat began to self-mutilate the left hind foot and toes (highlighted in the enlarged image on the right) within 3 days, despite preventive application of metronidazole/New Skin®. The animal was inhibited from doing further damage by the application of a bitter-tasting (Grannick's Bitter Apple®) spray. (Images courtesy of University of Pennsylvania, ULAR.)

resulted in limb injury, then degree of lameness, amount of swelling, and integrity of the wound should be assessed (Geertsema and Lindsell, 2011). Treatments should include analgesia, cleansing of the wound site, bandaging of the area following application of local analgesics and triple antibiotic ointments, and placement of a restraint collar to prevent the rat from accessing the injured area (see Chapter 4, "Restraint Collar Considerations"). Metronidazole can be applied ("painted") topically over the area where self-injury has occurred due to its aversive taste; a NewSkin® bandage can then be painted over the metronidazole to prolong the presence of the drug on the skin and promote healing (Zhang et al., 2001). If the self-injury is severe to the point of severe welfare alterations, euthanasia is recommended.

More often than not, moribund and paralyzed animals will require euthanasia if there is no improvement or change in activity status within 24 h of initial presentation.

Ocular Lesions Secondary to Anesthesia

- **Cause and impact:** Cloudiness of corneal tissue can be secondary to application of anesthesia and omission of topical eye lubrication while animals are under anesthesia. Corneal lesions and keratoconjunctivitis sicca (Kufoy et al., 1989) can be more severe in animals anesthetized with ketamine plus xylazine; minimal ocular changes have been noted in rats following enflurane or isoflurane anesthesia. Corneal lesions can be observed within 24 h after injectable anesthetic administration and may be irreversible. Compared with Sprague-Dawley and Lewis rats, Wistar, Long-Evans, and Fischer 344 rats had increased incidence and severity of corneal lesions after anesthesia with ketamine plus xylazine, suggesting that these three strains are at increased risk for developing postanesthetic corneal lesions with this regimen (Turner and Albassam, 2005).

  Acute reversible corneal lesions, attributable to a side effect of xylazine, have been documented in rats (Calderone et al., 1986).

- **Potential treatments:** Treatment with a sterile ophthalmic lubricant (e.g., Puralube™) can assist with the prevention of dry eyes and will soothe irritation. Ophthalmic products of this nature should be applied any time a rodent is under anesthesia to provide a protective film over the ocular surface, similar to what is done in routine veterinary clinical practice. The severity of corneal changes has been diminished in rats for which ketamine plus xylazine anesthesia was reversed with yohimbine (Turner and Albassam, 2005).

  Topical ophthalmic ointment, with or without added antibiotics, is warranted as a first-line approach even for animals that appear to be otherwise behaviorally normal, despite the ocular lesion. Certain lesions will be painful, with notation of animals scratching at the eye and face; these animals should receive topical (proparacaine) drops and systemic analgesics (meloxicam 5 mg/kg SC) for pain relief. Surgical enucleation may be warranted for correction of severe ocular injuries in rats.

Poor Body Condition

- **Cause and impact:** Rats may present clinically with a thin, alopecic, and hunched appearance without much forewarning (Figure 3.13). This may be attributed to malocclusion (see

Fig. 3.13 Rats may present clinically with a thin, alopecic, and hunched appearance. Animals should be treated with supportive care, and further diagnostics are warranted. (Image courtesy of University of Pennsylvania, ULAR.)

relevant section on this topic) that is preventing ingestion of hard food items (e.g., pelleted chow) or as a result of a gastrointestinal abnormality, husbandry alterations, experimental manipulation, or internal tumor burden (Mexas et al., 2011).

Differentials should include behavioral stereotypies that are preventing the animal from grooming or causing the animal to overgroom. Infectious agents should be ruled out as a cause of alopecia through the performance of skin scrapes and fungal cultures. If experimental treatments are potentially toxic or unpalatable, this should be further discussed with the research team.

- **Potential treatments:** The logical and prioritized causes should be treated first, typically including administration of a subcutaneous bolus of fluids, nutritional support, and heat supplementation if the animal is hypothermic. If the animals were expected to succumb to experimental disease, then aggressive treatment efforts should be aimed at assisting with nutritional supplements and comforting the animal with provision of cage enrichments and bedding substrates to attempt to reach the experimental endpoint.

  The rest of the affiliated colony should be evaluated further to determine if the condition is endemic, and blood sampling can be done for further diagnostic assessments of potential infectious pathogens. Body condition scoring should be monitored daily and BWs checked routinely to track any further decline. Animals that become quiet, less alert, and unresponsive should be considered for euthanasia prior to a moribund state and spontaneous death.

Ringtail

- **Cause and impact:** Ringtail is a pathologic condition of the tail, and sometimes feet, characterized by dry skin and annular constrictions that can result in necrosis and loss of portions of the tail (Figure 3.14).

  The cause of ringtail is not completely understood, although the condition is typically noted in weanling animals and may be caused by relative environmental humidity levels below 25%. Other contributing factors, such as dietary deficiencies, genetic susceptibility, environmental temperatures, and degree of hydration, also have been proposed. The variety of possible etiologic factors suggests that this syndrome might be the clinical expression of more than one causative agent or that more than one causative agent may be necessary to induce ringtail (Crippa et al., 2000).
- **Potential treatments:** Treatment with over-the-counter lanolin ointment (a nontoxic, inexpensive, and effective moisturizer) has been successful when initiated prior to the condition becoming severe enough that there is tail necrosis. It can also be applied prophylactically to rats starting at 7 days of age for groups that may have a history of disease (Taylor et al., 2006).

Ulcerative Dermatitis

- **Causes and impact:** Ulcerative dermatitis (UD) has been noted in Zucker lean rats, especially distal to forelimbs, with isolated lesions on the head and behind the ears. Determining the appropriate sensitivity profile to cultured bacteria is essential to provide an effective antibiotic treatment. Skin lesions may be secondary to dietary deficiencies, such as linoleic acid deficiency, with manifestation of focal areas of alopecia to diffuse areas of moist dermatitis on the head, face, ear pinnae, and neckline.
- **Potential treatments:** Administration of leptin topically at 5 μg daily to affected areas can provide reduction in wound size and severity. As well, trimming of hind toenails to prevent self-inflicted skin trauma is advisable (Oppelt, 2005).

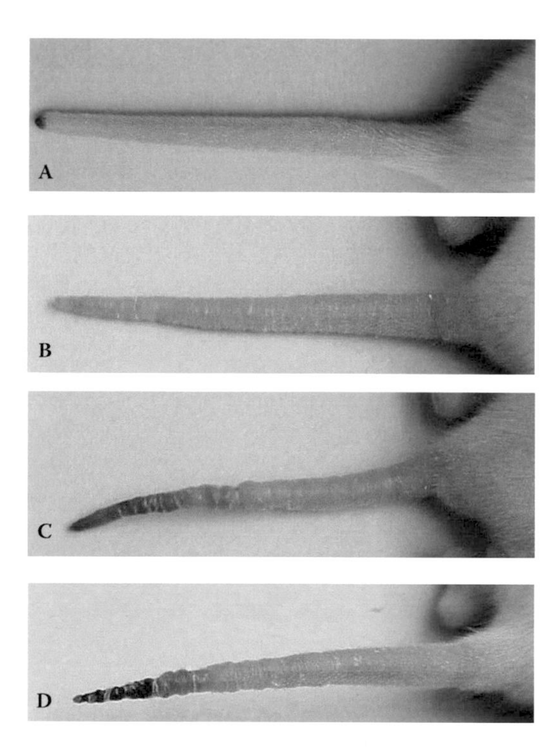

Fig. 3.14 Examples of ringtail in preweanling rats. (A) Normal, healthy rat pup tail. This tail was given a condition score of 0. Note that the distal portion of the tail has been biopsied for genotyping. Representative clinical cases of ringtail of varying severity: (B) Rat pup tail showing some flaking of the skin with mild constrictions. This tail was given a condition score of 1. (C) Rat pup tail clearly exhibiting annular constrictions and some malformation of tail tissue. This tail was given a condition score of 2. (D) Rat pup tail exhibiting annular constrictions with some malformation of tissue; the tail tip appears necrotic. This tail was given a condition score of 3. Topical application of lanolin to tails appearing like those in Figure 3.14B and 3.14C returned them to a healthy and clinical normal tail appearance. Due to the level of necrosis in Figure 3.14D, one might consider amputation of the tail tip to remove necrotic tissue. (Reprinted with permission from AALAS. Taylor, DK, Rogers, MM, and Hankenson, FC. 2006. *J Am Assoc Lab Anim Sci* 45:83–87.)

Concurrent correction of dietary imbalances, topical application of betadine cleanses, triple antibiotic ointment, and zinc oxide may be beneficial (Godfrey et al., 2005).

For a comprehensive listing of various treatments of UD in laboratory mice that may be efficacious for UD in laboratory rats, refer to the relevant section in Chapter 2.

Urolithiasis

- **Cause and impact:** Clinical signs indicative of urolithiasis include combinations of hematuria, red-stained bedding with abnormal urine, red-stained or wet pelage (especially over the abdomen), sensitivity to touch in the abdominal area, swollen or palpable kidney or bladder, unkempt fur, anorexia, reduced urination, reduced water intake, and unexpected weight loss or gain (due to fluid retention) (Newland et al., 2005).

  Partial-to-complete obstruction of urinary outflow can cause mild-to-severe pressure necrosis of the renal pelvis, medulla, and eventually the cortex. In addition, urinary calculi can inflame and cause degeneration and necrosis of the epithelial lining of the urinary tract. Incomplete emptying of the urinary system due to obstruction, coupled with the loss of epithelial integrity, allows bacterial overgrowth and subsequently an ascending urinary tract infection. In the case of a severe infection, bacteria can gain access to systemic circulation and cause sepsis.

  Urolithiasis has also been linked to a model of lymphocytic choriomeningitis virus (LCMV) infection in Lewis rats (Mook et al., 2004).

- **Potential treatments:** These factors indicated above, when taken as a whole, make it clear that once potentially obstructive uroliths form, the future health of the rat is at considerable risk, perhaps irreversibly, because calculi are highly persistent.

  Diet may need to be altered if the rats are to be maintained in the research colony. For example, of those rats maintained on a purified American Institute of Nutrition (AIN)-93 diet, males are considerably more at risk for urolithiasis and develop the condition within a few months of eating the diet (Newland et al., 2005). As rats on the AIN-93 diet aged, the discrepancy in risk between males and females increased; in fact, by 100 weeks, nearly 60% of male rats died of urolithiasis,

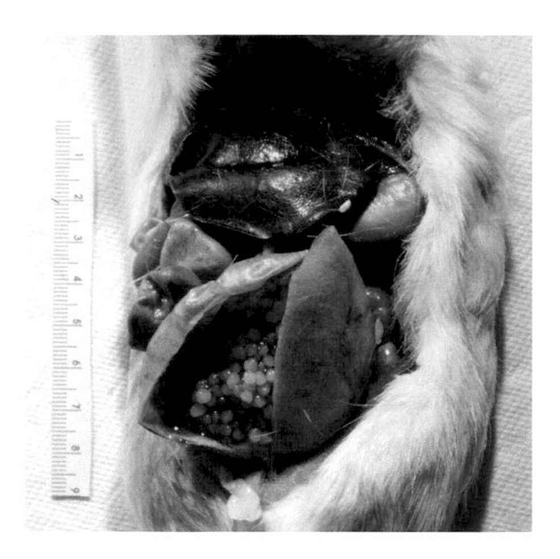

Fig. 3.15 Representation of severe urolithiasis in an adult Lewis rat experimentally infected with lymphocytic choriomeningitis virus. Bladder walls were markedly thick, and the bladder was enlarged and filled with multiple small uroliths. (Reprinted with permission from AALAS. Mook, DM, Painter, JA, Pullium, JK, Ford, TR, Dillehay, DL, and Pearce, BD. 2004. *Comp Med* 54:318–323.)

three times the prevalence seen in female rats. Postmortem analyses suggested that males were more likely to have bladder calculi than were females, who usually formed calculi in the kidney. Euthanasia of rats with severe clinical signs of urinary dysfunction, likely secondary to stone formation, is warranted to limit continued discomfort and potential spontaneous mortality (Figure 3.15).

research-related medical issues

Additional topics concerning laboratory rat health felt to warrant further information herein due to their prevalence in contemporary research environments are provided next in alphabetical order. Under each topic, "background information" is provided, and "potential treatments" offer suggestions about procedures, therapeutic treatments, and further considerations.

Arthritis Models

- **Background:** Induction of arthritis to better investigate the pathogenesis of inflammation and test the potential for antiarthritic agents is a classic model in the laboratory rodent. Differing models include adjuvant arthritis (typically in male Lewis rats, with injection at the tail base or into the foot pad); type II collagen arthritis (typically in female rats given bovine type II collagen); antigen arthritis; and injection of substances like capsaicin and carrageenan (Bendele, 2001). Tail and paw swelling with edema is expected as an acute inflammatory reaction, and the severity of paw swelling may render the animal immobile. While experimental treatments can be administered, these rats would typically be scientifically justified not to receive analgesics for pain management due to concern for impact on the development of experimental inflammation.

- **Potential treatments:** Provision of soft bedding and feeding of a softened food or nutritional supplement on the cage floor are recommended (Flecknell, 2001). Positioning of the water source, such that the animal can access fluids easily without additional pressure on the inflamed joints, should be considered. If analgesic treatments can be administered and not compromise the data, one could provide NSAIDs (indomethacin), dexamethasone or other corticosteroids (at low doses to avoid toxicity seen with chronic use), methotrexate (low dose), and biological agents like soluble tumor necrosis factor R2 (TNF-R2) that are currently marked for human treatments of arthritis (Bendele, 2001). It is not recommended that animals be handled daily as this may increase their stress and alter the desired inflammatory outcomes; instead, every-other-day (EOD) handling should be sufficient (Brand, 2005). Humane endpoints must be established for this type of model to best limit the duration and intensity of pain sensation (National Research Council [NRC], 2009).

Cranial Implant Maintenance

- **Background:** Neuroscience research often requires surgical implantation of an apparatus that permits direct manipulation of brain tissue or measurement of neuronal activity in conscious animals. Successful factors for longevity of cranial implants in rats have been described (Gardiner and Toth, 1999). Contributing factors include accurate targeting of the location of interest, aseptic surgical technique, maximal adherence of acrylic cement to the bone through proper preparation of the skull surface, and provision of ventilation during the thermogenic phase of cement curing. For the skin to heal properly around the implant site, apposition of skin to the implant is essential to promote comfort and reduce the likelihood of secondary bacterial infections.

- **Potential treatments:** Wound margins should be treated topically and liberally with antibiotic ointment (daily for 7 days postsurgery) (Gardiner and Toth, 1999). Repositioning of the skin to adhere better to the headpiece may be of use for a nonhealing incision site, and skin retraction may need to be employed to gain the coverage needed over the skull-cap. Systemic antibiotics can also be administered pre- and postsurgically.

Continued maintenance of the areas around the implant will likely be necessary to avoid the buildup of crusts and potential for secondary bacterial infections (Figure 3.16). Using a nonirritating antiseptic/cleansing solution, gently remove any scabbing and minimize disruption of wound margins. Trimming of hairs along the margin is useful to minimize irritation. Antibiotic ointment may then be applied as needed. Cultures should be routinely taken to track potential bacterial infections and provide relevant systemic antibiotics, if necessary.

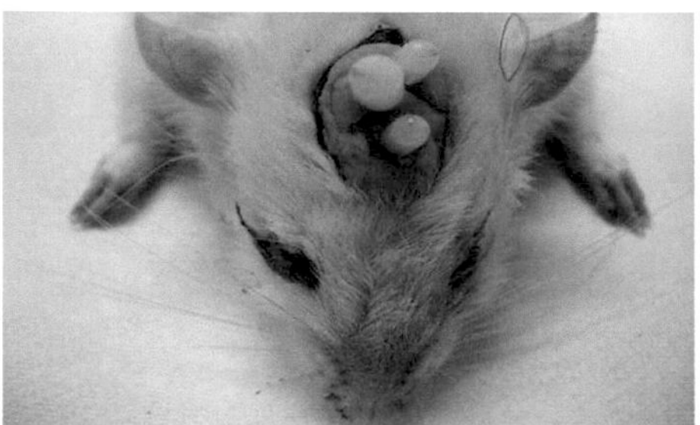

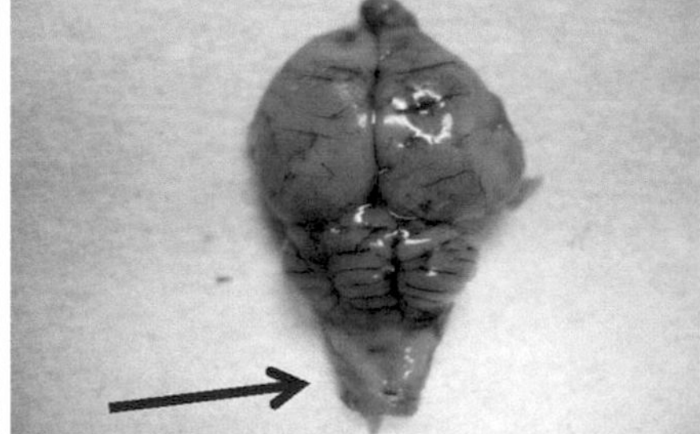

Fig. 3.16 Cranial implant complications in an adult rat. Porphyrin staining was noted around the eyes and nares, indicative of poor health and stress. Although the wound margins appeared relatively healthy, apposition was compromised. Due to declining health, this animal was euthanized, and *Pseudomonas aeruginosa* was cultured from the brainstem surface (arrow). (Images courtesy of University of Pennsylvania, ULAR.)

Incontinence Secondary to Spinal Cord Injury Models

- **Background:** Urine scald secondary to urinary incontinence from spinal cord injury studies can pose significant clinical problems. Urine scald is likely to cause discomfort, signified by skin with severe redness and warmth and the presence of urine or urine stains. Related complications can include intractable skin ulceration, secondary bacterial dermatitis, and self-trauma.
- **Potential treatments:** Amelioration of the discomfort has been attempted by application of a commercially available hexamethyldisiloxane (HMDS)-based skin protectant barrier film, 3M No-Sting Barrier Film® (3M Corporation, St. Paul, MN), which is used to treat diaper rash in human infants and urine scald in incontinent adults (St. Claire et al., 1997).

  For incontinent rats, urinary bladders can be manually expressed every 8 h until rats are observed to urinate without assistance. After each timed expression of the urinary bladder, barrier film is applied to clean, dry skin by spraying a uniform coat of film over the affected area. Animals should be observed for signs of discomfort after application of the barrier film. In addition, daily monitoring for paralysis, signs of dehydration, food intake, and evidence and degree of urine scald should be performed. Skin irritation can be rated from minor (slight redness, cool to touch) to major (severe redness, warm to touch) with or without moisture from urine (St. Claire et al., 1997).

  Treatment of spinal cord injury with minocycline, an antibiotic with neuroprotective effects, has been beneficial for restoration of motor coordination and hind limb reflexes. This antibiotic can be administered within 1 h after injury (90 mg/kg IP), followed by doses (45 mg/kg IP) twice daily for 5 days (Teng et al., 2004).

Middle Cerebral Artery Occlusion in Rat Models of Stroke

- **Background:** The major complication of the stroke model is the substantial morbidity and mortality that occurs postoperatively due to respiratory distress caused by stimulation of the sympathetic nerve system. Prolonged occlusion of the common carotid arteries can lead to prolonged tachycardia and potential for arrhythmias.

- **Potential treatments:** Administration of bupivacaine (0.25% solution 0.1–0.2 ml SC) can ameliorate left-side heart failure that would otherwise lead to mortality and may improve the outcome of the model (Wang Fischer et al., 2003). Scoring sheets have been developed for evaluation of pain management in this model as the administration of NSAIDs for pain relief may not be permitted due to bias of data outcomes (Kirsch et al., 2002).

Obese and Diabetic Rat Models

- **Background:** Rat models for obesity and diabetes research are beneficial to the identification of surgical and therapeutic interventions that can be translated to the related human disease syndromes. Performing surgery in physiologically compromised rats, particularly with their body conformation of a thinner thoracic cavity and larger abdominal mass, can result in adverse effects like hypoglycemia, difficulty with dehiscence of incision sites, and anesthesia reactions that can potentially lead to fatal outcomes.
- **Potential treatments:** For those rats undergoing surgery, attention to the minimal time for fasting, both presurgically and postsurgically, is key (see Chapter 4, "Fasting Considerations"). To prevent dehiscence of surgical sites, it is recommended to avoid surgical skin clips for this model and instead close incisions with continuous suture patterns, with minimal suture size, and subcuticular closure patterns. Maintaining the anesthetic dose of isoflurane to no more than 1.5%, with an oxygen flow rate of 0.5 L/min, facilitates recovery issues and maintains animals at a reasonable depth of anesthesia. As well, supplemental heat should be provided as described and endorsed for all surgical models in rodents. Application of these refinements has been shown to contribute to a survival rate of approximately 90% for gastrointestinal procedures performed in obese and diabetic rats (Baran et al., 2011).

  Providing supportive care and specialized environments will be best for these obese models. For animals with evidence of diabetes, more frequent cage changes and provision of more absorbent bedding substrates should be used in the rat cages to compensate for increased urine production.

Opportunistic Infections in Immunodeficient Rat Models

- **Background:** Similar to cases in laboratory mice that are immunodeficient (see relevant section in Chapter 2), *Klebsiella oxytoca* has been identified as a monoculture from urogenital tract infections and abscesses, as well as serving as the etiology for otitis, keratoconjunctivitis, meningitis, lymphadenitis, and pneumonia (Bleich et al., 2008). Abnormal colonization with *K. pneumonia* has also been documented following antibiotic treatment in nude rats (Hansen, 1995). Rats enrolled in longevity studies may succumb to opportunistic infections with age, prior to collection of desired data points.
- **Potential treatments:** Husbandry and environmental changes have been useful in eradicating opportunists, like *Pneumocystis carini*. Housing rats in autoclaved cages, with autoclaved bedding, and provision of trimethorim-sulfamethoxazole-treated acidified water have minimized reported heath issues. As well, providing a diet with 14% protein and 3.5% fat, along with pair housing of rats, has effectively extended the lifespan and improved overall health in aged rats (Zahorsky-Reeves et al., 2007). Also, nutritional supplementation in the form of sterile solidified gels may be provided (see Chapter 4, "Nutritional Therapy Considerations").

Pododermatitis

- **Background:** Pododermatitis can be common in mature rats (>300 g) chronically housed (>1 year) in wire-bottom cages but is less commonly noted when animals are housed on bedding (Carraway and Witt, 2003, Peace et al., 2001). The problem is characterized by chronic, suppurative inflammatory lesions (ulcers) on the plantar surfaces of the hind feet; lesions may be reddened and raised, with keratinized growth developing into crusts and scabs (Peace et al., 2001, Sharp and Villano, 2012).
- **Potential treatments:** Topical and systemic treatment options may be limited by impacts on study data; however, antibiotics and analgesics would be ideal for addressing the infectious nature and associated pain from these lesions. Placement of some sort of flattened and softer bedding substrate or surface on the wire cage bottom, akin to sterile gauze squares (4 x 4 inches), has a significant preventive benefit for diminishing the potential for ulceration of noted foot sores

(Dimeo and Mitchell, 2005, Peace et al., 2001). Soaking of affected feet (hydrotherapy) in Epsom salt solution (4 cups of warm water to 1 teaspoon of salt) has anecdotally been successful for resolution of surface infection and softening of crusts covering foot wounds. Surgical debridement is rarely successful, and prognosis for complete resolution is guarded (Langlois, 2004). Closely monitor affected animals to ensure that any rats experiencing severe pain and distress are removed from the study and euthanized.

Spontaneously Hypertensive Rat Models

- **Background:** To promote maintenance of BW of senescent female spontaneously hypertensive rats (SHRs), supplementing powdered feed is useful to offset loss of appetite and weight loss.
- **Potential treatments:** With age, SHR rats will benefit from the addition of powdered food to ensure that BWs remain stable and to prevent malnutrition that could lead to premature death. Rats were also given powdered rat chow in shallow bowls to facilitate the eating and digestion of food. As the female SHR matures, special care and handling are essential to help maintain BW and good health. With only modest changes in routine (i.e., powdered food) and an attentive eye on the rats' daily activities, it is possible to maintain these rats in a healthy condition until the termination of the study (Belanger et al., 1999).

Tumor Burden in Rat Models

- **Background:** Rodent tumor models are quite common in laboratory animal facilities. Institutional guidance should be followed with respect to size of allowable tumors and increased monitoring of animal health (see Chapter 4, "Tumor Development and Monitoring Considerations"). Spontaneously occurring tumors may also develop and should be managed based on how the animal's overall health and body condition fares.

Tumors in rats may be secondary to foreign body reactions, particularly for intra-abdominal telemetry devices in certain strains (Popovic et al., 2004). Tumor incidence should be considered an adverse outcome in instrumented rats.

Subcutaneous masses involving the mammary chain are usually benign fibroadenomas, with less than 10% being malignant. Mammary tissue in rats is extensive, and masses can occur anywhere from the neck to the inguinal region, arising in locations as dorsal as the flank areas and across the shoulders.

Paraneoplastic syndrome in young rats has been described secondary to extensive mammary neoplasia (Figure 3.17) (Mexas et al., 2011).

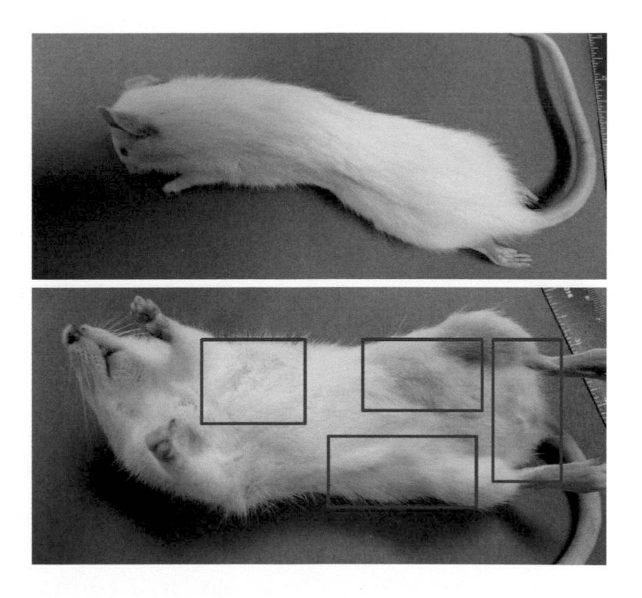

Fig. 3.17 Induced neoplasia model young (2-month-old) female rat. (Top) The animal was very thin with a BCS on initial examination at 1.5 of 5. (Bottom) Visible and firm palpable masses extended bilaterally throughout regions of mammary tissue on the ventral aspect of the mouse (highlighted in boxes). The animal had a 48-h history of lethargy and dehydration; on physical examination, the rat became extremely stressed, developed agonal breathing, and was euthanized immediately. Necropsy identified widespread mammary tumors (corresponding with palpated masses) and multiple organ abnormalities, including calcification as a paraneoplastic syndrome. (Images courtesy of University of Pennsylvania, ULAR.)

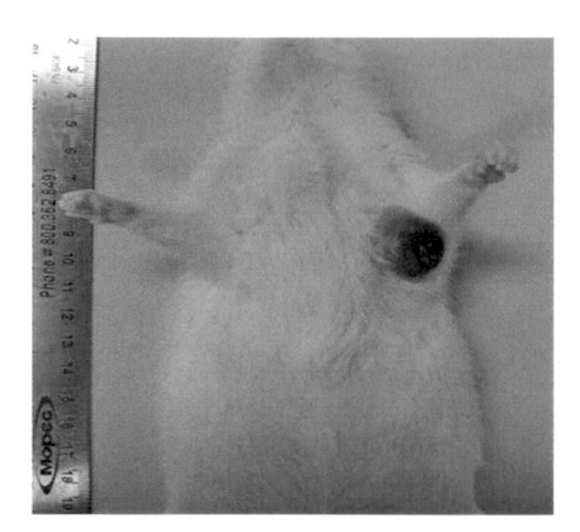

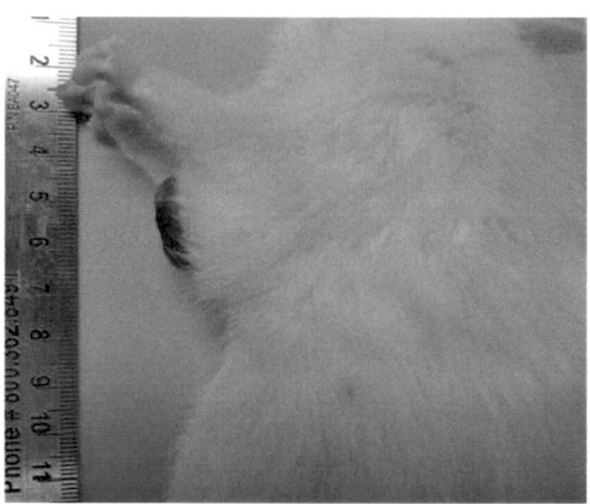

Fig. 3.18 Rat with a spontaneous ulcerated mammary tumor. Due to the ulceration and location in the left axillary region (left, ventral view; right, right-side recumbent view), this rat would require heightened monitoring for alterations to mobility, hemorrhage of the mass, and further irritation to the tumor site. (Images courtesy of University of Pennsylvania, ULAR.)

- **Potential treatments:** Surgical excision is the most common form of therapy and results in more cures than all other modalities combined (Mehler et al., 2004). Mammary gland tumor removal can be straightforward; in brief, the vascular supply to these tumors is limited and therefore can be ligated using vascular clamps or suture material. Once the neoplastic tissue is removed, the tissue space and subcutaneous tissue can be closed with 3–0 vicryl suture (using a simple-interrupted or continuous pattern). Overlying skin can be closed with suture, wound clips, or tissue glue (Fisher, 2002).

  It may be possible to have the tumor treated in some other manner to continue using the animal in a study; however, overall animal welfare should not be compromised if the tumor is left untreated, affects mobility, or ulcerates (Figure 3.18). Tumor development may affect animal welfare for those animals in long-term studies and decrease confidence in the reliability of data outcomes from the model.

euthanasia

Euthanasia is the process of inducing painless death in animals. To the greatest extent possible, animals being euthanized should not experience pain, fear, or other significant stress prior to their death.

Carbon dioxide (CO_2) exposure or narcosis is a frequently used euthanasia method for small laboratory animals due to its rapid onset of action, safety, low cost, and ready availability. Exposure times for carbon dioxide differ dramatically depending on the age of the rat to be euthanized; rats older than 21 days typically require 5 min of exposure time (Pritchett-Corning, 2009). Injectable and inhalant methods are therefore preferred unless individuals have received hands-on training for physical methods of euthanasia. Further discussion is provided in Chapter 4, "Euthanasia Considerations" and in the *AVMA Guidelines for the Euthanasia of Animals* (American Veterinary Medical Association [AVMA], 2013).

references

American Veterinary Medical Association (AVMA). 2013. *AVMA Guidelines for the Euthanasia of Animals:* 2013 Edition, pp. 1–102. https://www.avma.org/KB/Policies/Documents/euthanasia.pdf.

Arzadon, J. 2011. Adaptation of submandibular blood collection in rats. *J Am Assoc Lab Anim Sci* 50:750–751.

Atcha, Z, Rourke, C, Neo, AH, Goh, CW, Lim, JS, Aw, CC, Browne, ER, and Pemberton, DJ. 2010. Alternative method of oral dosing for rats. *J Am Assoc Lab Anim Sci* 49:335–343.

Banks, RE, Sharp, JM, Doss, SD, and Vanderford, DA. 2010. Rats, pp. 81–92. In *Exotic Small Mammal Care and Husbandry*. Wiley-Blackwell, Ames, IA.

Baran, SW, Loughery, CA, Zhuang, R, Maldonado, V, and Lin, DC. 2011. Refinement of gastrointestinal procedures in rat models for obesity and diabetes studies. *J Am Assoc Lab Anim Sci* 50:744.

Belanger, MP, Wallen, WJ, and Wittnich, C. 1999. Special feeding and care of senescent spontaneously hypertensive rats. *Contemp Top Lab Anim Sci* 38:7–11.

Bendele, A. 2001. Animal models of rheumatoid arthritis. *J Musculoskelet Neuronal Interact* 1:377–385.

Bleich, A, Kirsch, P, Sahly, H, Fahey, J, Smoczek, A, Hedrich, HJ, and Sundberg, JP. 2008. *Klebsiella oxytoca*: opportunistic infections in laboratory rodents. *Lab Anim* 42:369–375.

Brand, DD. 2005. Rodent models of rheumatoid arthritis. *Comp Med* 55:114–122.

Calderone, L, Grimes, P, and Shalev, M. 1986. Acute reversible cataract induced by xylazine and by ketamine-xylazine anesthesia in rats and mice. *Exp Eye Res* 42:331–337.

Caro, A, Brice, AK, and Veeder, CL. 2011. Investigation into a fire caused by improper surgical preparation and surgical instrumentation use in a protocol-related rodent surgery. *J Am Assoc Lab Anim Sci* 50:745.

Carraway, JH, and Witt, WM. 2003. Atypical occurrence of pododermatitis in Sprague-Dawley rats housed on hardwood chip bedding. *Contemp Top Lab Anim Sci* 42:109.

Ceccarelli, AV, and Rozengurt, N. 2002. Outbreak of hind limb paralysis in young CFW Swiss Webster mice. *Comp Med* 52:171–175.

Cloutier, S, Wahl, K, and Newberry, RC. 2010. Playful handling mitigates the stressfulness of injections in laboratory rats. *J Am Assoc Lab Anim Sci* 49:721.

Coria-Avila, GA, Gavrila, AM, Menard, S, Ismail, N, and Pfaus, JG. 2007. Cecum location in rats and the implications for intraperitoneal injections. *Lab Anim (NY)* 36:25–30.

Crippa, L, Gobbi, A, Ceruti, RM, Clifford, CB, Remuzzi, A, and Scanziani, E. 2000. Ringtail in suckling Munich Wistar Fromter rats: a histopathologic study. *Comp Med* 50:536–539.

Dimeo, D, and Mitchell, S. 2005. Foot sore development comparison data and the use of gauze for rats housed in wire-bottomed caging. *Contemp Top Lab Anim Sci* 44:65–66.

Dontas, IA, Tsolakis, AI, Khaldi, L, Patra, E, and Lyritis, GP. 2010. Malocclusion in aging Wistar rats. *J Am Assoc Lab Anim Sci* 49:22–26.

Fisher, PG. 2002. Surgical removal of rat mammary tumors. *Exotic DVM* 4.2:6.

Fitzner Toft, M, Petersen, MH, Dragsted, N, and Hansen, AK. 2006. The impact of different blood sampling methods on laboratory rats under different types of anaesthesia. *Lab Anim* 40:261–274.

Flecknell, PA. 2001. Analgesia of small mammals. *Vet Clin North Am Exot Anim Pract* 4:47–56, vi.

Flecknell, PA, Orr, HE, Roughan, JV, and Stewart, R. 1999. Comparison of the effects of oral or subcutaneous carprofen or ketoprofen in rats undergoing laparotomy. *Vet Rec* 144:65–67.

Flecknell, PA, Roughan, JV, and Stewart, R. 1999. Use of oral buprenorphine ("buprenorphine Jello") for postoperative analgesia in rats—a clinical trial. *Lab Anim* 33:169–174.

Gardiner, TW, and Toth, LA. 1999. Stereotactic surgery and long-term maintenance of cranial implants in research animals. *Contemp Top Lab Anim Sci* 38:56–63.

Geertsema, R, and Lindsell, C. 2011. Clinical scoring system for the evaluation of autophagia in laboratory rats. *J Am Assoc Lab Anim Sci* 50:819.

Godfrey, DM, Gaumond, GA, Delano, ML, and Silverman, J. 2005. Clinical linoleic acid deficiency in Dahl salt-sensitive (SS/Jr) rats. *Comp Med* 55:470–475.

Hansen, AK. 1995. Antibiotic treatment of nude rats and its impact on the aerobic bacterial flora. *Lab Anim* 29:37–44.

Hickman, DL, and Swan, M. 2010. Use of a body condition score technique to assess health status in a rat model of polycystic kidney disease. *J Am Assoc Lab Anim Sci* 49:155–159.

Kirsch, JH, Klaus, JA, Blizzard, KK, Hurn, PD, and Murphy, SJ. 2002. Pain evaluation and response to buprenorphine in rats subjected to sham middle cerebral artery occlusion. *Contemp Top Lab Anim Sci* 41:9–14.

Klaphake, E. 2006. Common rodent procedures. *Vet Clin North Am Exot Anim Pract* 9:389–413, vii–viii.

Koch, A, Scorpio, DG, and Ruben, D. 2008. Evaluation of supplemental warming methods for prevention of hypothermia in rats. *J Am Assoc Lab Anim Sci* 47:100–101.

Kohlert, DJ. 2012. Unanesthetized sublingual blood collection in rats. *J Am Assoc Lab Anim Sci* 51:658.

Kohn, DF, Martin, TE, Foley, PL, Morris, TH, Swindle, MM, Vogler, GA, and Wixson, SK. 2007. Public statement: guidelines for the assessment and management of pain in rodents and rabbits. *J Am Assoc Lab Anim Sci* 46:97–108.

Kufoy, EA, Pakalnis, VA, Parks, CD, Wells, A, Yang, CH, and Fox, A. 1989. Keratoconjunctivitis sicca with associated secondary uveitis elicited in rats after systemic xylazine/ketamine anesthesia. *Exp Eye Res* 49:861–871.

Kulick, LJ, Clemons, DJ, Hall, RL, and Koch, MA. 2005. Refinement of the urine concentration test in rats. *Contemp Top Lab Anim Sci* 44:46–49.

Kurien, BT, Everds, NE, and Scofield, RH. 2004. Experimental animal urine collection: a review. *Lab Anim* 38:333–361.

Langford, DJ, Bailey, AL, Chanda, ML, Clarke, SE, Drummond, TE, Echols, S, Glick, S, Ingrao, J, Klassen-Ross, T, Lacroix-Fralish,

ML, Matsumiya, L, Sorge, RE, Sotocinal, SG, Tabaka, JM, Wong, D, van den Maagdenberg, AM, Ferrari, MD, Craig, KD, and Mogil, JS. 2010. Coding of facial expressions of pain in the laboratory mouse. *Nat Methods* 7:447–449.

Langlois, I. 2004. Wound management in rodents. *Vet Clin North Am Exot Anim Pract* 7:141–167.

Liu, JY, Diaz, TG, 3rd, Vadgama, JV, and Henry, JP. 1996. Tail sectioning: a rapid and simple method for repeated blood sampling of the rat for corticosterone determination. *Lab Anim Sci* 46:243–245.

Luo, Y, Corning, BF, White, WJ, Fisher, TF, and Morin, RR. 2003. An evaluation of a rodent water replacement source. *Contemp Top Lab Anim Sci* 42:119.

Luo, Y, Nelson, RM, Luo, Y, Corning, BF, White, WJ, and Fisher, TF. 2004. An evaluation of chronic venous catheter patency. *Contemp Top Lab Anim Sci* 43:77.

Martin, LB, Thompson, AC, Martin, T, and Kristal, MB. 2001. Analgesic efficacy of orally administered buprenorphine in rats. *Comp Med* 51:43–48.

Mathieu, C. 2011. Comparative analysis of blood sampling techniques in the rat. *J Am Assoc Lab Anim Sci* 50:738.

Mehler, SJ, and Bennett, RA. 2004. Surgical oncology of exotic animals. *Vet Clin North Am Exot Anim Pract* 7:783–805, vii–viii.

Mexas, AM, Brice, AK, Young, B, Chodosh, LA, and Hankenson, FC. 2011. Paraneoplastic syndrome associated with severe multiorgan failure in a Lewis rat model of mammary neoplasia. *J Am Assoc Lab Anim Sci* 50:735.

Miller, AL, and Richardson, CA. 2011. Rodent analgesia. *Vet Clin North Am Exot Anim Pract* 14:81–92.

Mogil, JS. 2009. Animal models of pain: progress and challenges. *Nat Rev Neurosci* 10:283–294.

Molthen, RC. 2006. A simple, inexpensive, and effective light-carrying laryngoscopic blade for orotracheal intubation of rats. *J Am Assoc Lab Anim Sci* 45:88–93.

Mook, DM, Painter, JA, Pullium, JK, Ford, TR, Dillehay, DL, and Pearce, BD. 2004. Urolithiasis associated with experimental lymphocytic choriomeningitis virus inoculation in Lewis rats. *Comp Med* 54:318–323.

National Research Council (NRC). 2009. *Recognition and Alleviation of Pain in Laboratory Animals*. National Academies Press, Washington, DC.

Newland, MC, Reile, PA, Sartin, EA, Hart, M, Craig-Schmidt, MC, Mandel, I, and Mandel, N. 2005. Urolithiasis in rats consuming a dl bitartrate form of choline in a purified diet. *Comp Med* 55:354–367.

Nishijima, K, Kuwahara, S, Ohno, T, Miyaishi, O, Ito, Y, Makino, S, Sumi, Y, and Tanaka, S. 2007. Natural dental caries in molars of osteogenic disorder Shionogi rats. *Comp Med* 57:590–593.

Oppelt, KA and Salleng, K. 2005. An outbreak of ulcerative dermatitis in Zucker lean rats. *Contemp Top Lab Anim Sci* 44:70–71.

Ordodi, VL, Mic, FA, Mic, AA, Sandesc, D, and Paunescu, V. 2005. A simple device for intubation of rats. *Lab Anim (NY)* 34:37–39.

Ordodi, VL, Mic, FA, Mic, AA, Tanasie, G, Ionac, M, Sandesc, D, and Paunescu, V. 2006. Bone marrow aspiration from rats: a minimally invasive procedure. *Lab Anim (NY)* 35:41–44.

Paul-Murphy, J. 1996. Little critters: emergency medicine for small rodents, pp. 714–718. Fifth International Veterinary Emergency and Critical Care Symposium, San Antonio, TX.

Peace, TA, Singer, AW, Niemuth, NA, and Shaw, ME. 2001. Effects of caging type and animal source on the development of foot lesions in Sprague Dawley rats (*Rattus norvegicus*). *Contemp Top Lab Anim Sci* 40:17–21.

Popovic, A, Schenck, EJH, West, HA, and Swallow, JJ. 2004. Neoplasms in inbred and outbred rats with intra-abdominal and radiotelemetry devices. *Contemp Top Lab Anim Sci* 43:48.

Pritchett-Corning, KR. 2009. Euthanasia of neonatal rats with carbon dioxide. *J Am Assoc Lab Anim Sci* 48:23–27.

Rivera, B, Miller, S, Brown, E, and Price, R. 2005. A novel method for endotracheal intubation of mice and rats used in imaging studies. *Contemp Top Lab Anim Sci* 44:52–55.

Robinson, MA, Herron, AJ, Goodwin, BS, and Grill, RJ. 2012. Suprapubic bladder catheterization of male spinal-cord-injured Sprague-Dawley rats. *J Am Assoc Lab Anim Sci* 51:76–82.

Roughan, JV, and Flecknell, PA. 2004. Behaviour-based assessment of the duration of laparotomy-induced abdominal pain and the analgesic effects of carprofen and buprenorphine in rats. *Behav Pharmacol* 15:461–472.

Shapiro, R, Mason-Bright, T, and Mason, M. 2010. Implementation of an alternative intravenous injection method to reduce postoperative pain and distress in male rats. *J Am Assoc Lab Anim Sci* 49:685.

Sharp, PE, and LaRegina, MC. 1998. *The Laboratory Rat*. CRC Press, Boca Raton, FL.

Sharp, PE, and Villano, J. 2012. *The Laboratory Rat*, 2nd edition. CRC Press, Boca Raton, FL.

Sotocinal, SG, Sorge, RE, Zaloum, A, Tuttle, AH, Martin, LJ, Wieskopf, JS, Mapplebeck, JC, Wei, P, Zhan, S, Zhang, S, McDougall, JJ, King, OD, and Mogil, JS. 2011. The Rat Grimace Scale: a partially automated method for quantifying pain in the laboratory rat via facial expressions. *Mol Pain* 7:55.

St. Claire, MB, St. Claire, MC, Davis, JA, Chang, L, and Miller, GF. 1997. Barrier film protects skin of incontinent rats. *Contemp Top Lab Anim Sci* 36:46–48.

St. Claire, MB, Sowers, AL, Davis, JA, and Rhodes, LL. 1999. Urinary bladder catheterization of female mice and rats. *Contemp Top Lab Anim Sci* 38:78–79.

Taylor, DK. 2007. Study of two devices used to maintain normothermia in rats and mice during general anesthesia. *J Am Assoc Lab Anim Sci* 46:37–41.

Taylor, DK, Rogers, MM, and Hankenson, FC. 2006. Lanolin as a treatment option for ringtail in transgenic rats. *J Am Assoc Lab Anim Sci* 45:83–87.

Teng, YD, Choi, H, Onario, RC, Zhu, S, Desilets, FC, Lan, S, Woodard, EJ, Snyder, EY, Eichler, ME, and Friedlander, RM. 2004. Minocycline inhibits contusion-triggered mitochondrial cytochrome c release and mitigates functional deficits after spinal cord injury. *Proc Natl Acad Sci U S A* 101:3071–3076.

Toth, LA, and Gardiner, TW. 2000. Food and water restriction protocols: physiological and behavioral considerations. *Contemp Top Lab Anim Sci* 39:9–17.

Turner, PV, and Albassam, MA. 2005. Susceptibility of rats to corneal lesions after injectable anesthesia. *Comp Med* 55:175–182.

Turner, PV, Brabb, T, Pekow, C, and Vasbinder, MA. 2011. Administration of substances to laboratory animals: routes of administration and factors to consider. *J Am Assoc Lab Anim Sci* 50:600–613.

Turner, PV, Pekow, C, Vasbinder, MA, and Brabb, T. 2011. Administration of substances to laboratory animals: equipment considerations, vehicle selection, and solute preparation. *J Am Assoc Lab Anim Sci* 50:614–627.

Turner, PV, Vaughn, E, Sunohara-Neilson, J, Ovari, J, and Leri, F. 2012. Oral gavage in rats: animal welfare evaluation. *J Am Assoc Lab Anim Sci* 51:25–30.

Wachtman, LM, Browning, MD, Bedja, D, Pin, S, and Gabrielson, KL. 2006. Validation of the use of long-term indwelling jugular catheters in a rat model of cardiotoxicity. *J Am Assoc Lab Anim Sci* 45:55–64.

Wang Fischer, YL, Renzi, MJ, Hall, L, Wixson, SK, Thirumalai, N, Kirschner, T, Du, FY, Miller, JA, Jolliffe, LK, and Farrell, FX. 2003. Refined technique for inducing and grading middle cerebral artery occlusion in rat stroke model. *Contemp Top Lab Anim Sci* 42:78.

Weiner, CM, Morrissey, J, Kalbfliesh, M, Wilwol, M, Multari, H, and Smith, JC. 2012. Four-week evaluation of jugular vein catheter patency in rats using different maintenance regimens. *J Am Assoc Lab Anim Sci* 51:656.

Zahorsky-Reeves, J, Lawson, G, Chu, DK, Schimmel, A, Ezell, PC, Dang, M, and Couto, M. 2007. Maintaining longevity in a triple transgenic rat model of Alzheimer's disease. *J Am Assoc Lab Anim Sci* 46:124.

Zeleski, KL, Orr-Gonzalez, S, and Lambert, L. 2011. Go for the jugular! Blood draw refinement in rats. *J Am Assoc Lab Anim Sci* 50:752.

Zhang, YP, Onifer, SM, Burke, DA, and Shields, CB. 2001. A topical mixture for preventing, abolishing, and treating autophagia and self-mutilation in laboratory rats. *Contemp Top Lab Anim Sci* 40:35–36.

special considerations for critical care management in laboratory rodents

introduction

Critical care monitoring for small rodent patients is notoriously challenging, complicated further by physiologic idiosyncrasies and limitations on applicable treatments and interventions (Hawkins and Graham, 2007, Lichtenberger, 2007, Lichtenberger and Ko, 2007). This chapter offers important supplementary information for critical care for the laboratory rodent patient, with specifics on types of models with inherent potential for pain and distress. The comprehensive subject matter herein was consolidated from a variety of institutional guidance documents, publications, and abstracts to highlight particular regulatory, clinical, and experimental facets of working with these species. Due to the diversity of material covered here, the topics are in *alphabetical order*, akin to the listings of considerations in previous chapters, for ease of location.

In addition, readers are strongly encouraged to review their own institutional guidelines and policies on these topics to best address animal welfare and critical care management issues, ideally in consultation with available veterinary staff.

aging animal model considerations

For biogerontology research, animals with median life spans longer than 20 months are usually acceptable to achieve experimental objectives (Nadon, 2004). The most commonly used strain for aging research in mice is the C57BL/6, and the most commonly used rat strain is the F344. Aging is often accompanied by dysfunction or disease of bodily systems, manifesting as cardiovascular system disorders, renal and respiratory function decline, osteoarthritis, osteoporosis, cataracts, disease-promoting mutations, amyloidosis, leukemias, and other cancers (Dontas et al., 2010, Nadon, 2004). In aging mice, enlargement of preputial glands is common, particularly bilaterally. However, these mice can still be used in research (potentially excluding reproductive studies) because duct ectasia is painless and does not commonly require treatment (Donnelly and Walberg, 2011). An increased incidence of malocclusion has been noted in aging Wistar rats; therefore, attention to dental health, and provision of hard substances on which to gnaw, should be heightened in aging colonies (Dontas et al., 2010, Nadon, 2004). Older rodents may develop arthritis, which may lead to chronic pain; this can be treated with a course of anti-inflammatory medications (2–3 weeks), followed by a 1-week break, then resumption of analgesic treatment (Flecknell, 2001). Neoplasia is also more common in aged laboratory rodents, and specific strain idiosyncrasies have been reviewed elsewhere (Danneman et al., 2012, Nadon, 2004).

Although many of these disease syndromes are progressive, rarely are they life threatening; these aged animals can serve as unique models in which to evaluate age-related physiologic alterations. Unfortunately, the ability to detect subtle alterations due to aging and subclinical declines in health is challenging despite daily observations of laboratory rodents. Similar to objective scoring systems for monitoring health during experimental procedures, these types of scoring systems can also be utilized for evaluation of aging animals (Figure 4.1) (Phillips et al., 2010). Efforts have been undertaken to determine markers of imminent death in aging mice by looking at temperature and body weight changes over time (Ray et al., 2010, Trammell et al., 2012).

Drug therapy in aging animals should be evaluated with some caution. Older (geriatric) rodents may likely have varying degrees of organ dysfunction and altered drug metabolism. Hepatic and renal dysfunction can have a negative impact on attempts at drug therapy

Date: *May* 28, 2009	Rat ID	C49
Notes: *Teeth need checking Weekly*	Experiment	300
	Date of birth	2-27-0 7
	Age	27 *mo*
	Possible score	Actual score
Appearance: observe rat in cage		
Appearance is normal: no obvious skin or coat problems	1	
Slightly abnormal: slight change in coat or skin—could be dirty coat, dander, or inflammation	2	
Moderately abnormal: obvious change from normal—more dullness, dirty, skin inflammation	3	3
Severely abnormal: marked change from normal—skin and coat very dull, dirty, or inflamed	4	
Extremely abnormal: extreme signs of deterioration of coat and skin	5	
Posture: observe rat while still in the cage or walking		
Normal posture: sitting, standing, or rearing normally	1	
Slightly abnormal posture: slight flatness or hunch	2	
Moderately abnormal: possibly dragging belly or hunched	3	3
Severely abnormal: mostly flat with little elevation off surface	4	
Extremely abnormal: flat to surface, unable to elevate off floor of cage	5	
Moblity: observe rat's ability to move around the cage		
No impairment: able to move normally	1	
Slightly impaired: may have some ataxia or splay causing slight problems with movement	2	
Moderately impaired: obvious mobility problems	3	
Severely impaired: definite problems with being able to move easily about the cage	4	4
Extremely impaired: unable to move at all or without extreme hardship	5	
Muscle Tone: hold rat in hand to assess tone of rear legs and abdomen		
Normal muscle tone: muscle groups have normal tone or mass	1	
Slightly abnormal: muscle mass slightly soft	2	
Moderately abnormal: muscle mass less firm, abdomen slightly soft	3	3
Severely abnormal: muscle mass very thin, soft, undefined	4	
Extremely abnormal: muscle mass has no tone or definition	5	
Total Score	4–20	13
Current weight		350.0
Previous weight		352.0
Weight gain (+) or loss (–)		–2.0

Fig. 4.1 Sample score sheet for observational assessment, with brief descriptions of the four measures, for aged rats. For each measure, the highest (most abnormal) score possible is 5. The total composite score possible for all measures combined is 20. In addition, body weight for the current and previous weeks is included, along with whether a weight gain or loss was present. (Reprinted with permission from AALAS. Phillips, PM, Jarema, KA, Kurtz, DM, and MacPhail, RC. 2010. *J Am Assoc Lab Anim Sci* 49:792–799.)

because the liver and kidneys are involved in the function, breakdown, and clearance of most drugs. Dosing of certain medications may need to be lowered and administered with less frequency in aged patients. Renal disease can diminish the kidney's ability to filter and excrete drugs, leading to the prolonged presence of drug and drug metabolite in the body system. Therefore, drug types like aminoglycosides (e.g., gentamicin, amikacin), which are primarily removed by the kidney, should be used with caution. As well, administration of potassium-containing fluids could lead to life-threatening serum potassium levels in animals with renal disease.

If aged animals are overweight or obese, one should consider dosing at their "ideal" adult body weight, especially when using drugs that distribute to adipose tissues, such as thiobarbiturates. In general, animals with an abnormal body condition may have underlying medical problems that could affect drug choices; every effort should be made to identify the cause of the altered body condition before starting drug treatment.

blood loss considerations

Blood loss, secondary to hemorrhage, hemophilia, immune disorders, or experimental sampling, may lead to physiologic alterations that result in a critical health concern. An animal with chronic anemia may tolerate a loss of 60% of blood volume before a transfusion is necessary. Assessment of the patient is key to best determine if the animal is tolerating the anemia as blood transfusion has inherent risks (Morrisey, 2003b).

Telemetric monitoring has been used during repeated blood drawing to observe systemic effects of blood sampling/loss on mice. In these studies, measurement of electrocardiogram (ECG), body temperature, and blood pressure were performed while imitating a blood sampling loss of up to 40% of the animals' total blood volume (TBV). Results noted on the ECG included declines in heart rates after 30–40% of the TBV was withdrawn. In general, transfusions are warranted for acute blood losses of greater than 30% of an animal's blood volume or, alternatively, if there is a 50% decrease from a baseline packed cell volume level (Morrisey, 2003b). In a related study, mean blood pressure declined after 20–30% withdrawal of TBV. Body temperature showed declines after 20% blood withdrawal; therefore, these quantitative assessments support policies on limiting blood drawing to 20% of TBV (Mercogliano et al., 2002).

Whole blood can potentially be transfused as needed but should be transfused from like species to like species. Blood typing does not appear to be necessary for most small mammals, and transfusion reactions are rare. Premedication with diphenhydramine at 1 mg/kg (intravenously [IV] or intramuscularly [IM]) may be beneficial to limit adverse transfusion reactions (Mader, 2002).

chronic indwelling device considerations

Catheters and vascular access ports are implanted routinely in laboratory animal patients being treated for cardiovascular diseases. Management of device-associated infections is critical; development of secondary complications can pose a serious health risk, particularly if the animals are immunodeficient. In animals, infections associated with long-term catheterization not only are potentially detrimental to their well-being but also are likely to cause variation in the experimental data obtained from such animals.

Animals should be treated perioperatively with antibiotics when long-term catheters are placed, potentially for a period of 5 to 7 days. Diagnosis of infections, should they develop, may be based on clinical signs of infection, including purulent discharge from the catheter exit site, swelling or erythema at the exit site or along the subcutaneous catheter route, or abscess at any point along the catheter tract. Temperatures may be elevated in these animals, with a decrease in appetite and activity and high white blood cell counts on blood tests. Specimens from suspect infections should be submitted for bacterial culture and antibiotic sensitivity testing.

Localized treatment may be performed with removal of purulent material from the catheter tract and disinfection/flushing of the tract with a povidone-iodine containing solution, similar to what is done in other species (Taylor and Grady, 1998). If health concerns increase (continued fever, anorexia, lethargy), with failure to respond to topical disinfection and systemic antibiotics, the decision to remove the catheter should be made in the interest of animal welfare.

depilatory cream considerations

Chemical depilatory creams are often used for fur removal, prior to experimental procedures or surgery, in laboratory rodents. It may be desirable to remove hair over blood collection sites (saphenous vessels), ECG lead-positioning points, for imaging purposes, prior to tumor cell implantation, or for surgical field access. Depilatory creams for rodents work well, but creams should be completely wiped away with a cloth and warm sterile solutions to prevent inadvertent ingestion during grooming and potential for skin irritation/inflammation, which can confound research outcomes.

Animals should remain under anesthesia during the contact time with the depilatory. Depilatories include over-the-counter products like Nair® (with or without aloe for sensitive skin) and Veet®. They can be applied directly to the fur of the rodent or applied after clipping the fur. The product may be left on for several minutes (1–10 min) then wiped off with a sterile gauze pad or tissue; if necessary, the application process can be repeated until hair removal is complete (Angel et al., 1992, Finlay et al., 2012). No animal should be left unattended during the application of depilatory agents.

Manual hair removal is not advised as an alternative; instead, electric shavers or handheld razors should be employed with the animal under anesthesia.

equipment considerations for rodent surgery and emergency procedures

Prior to the presentation of critical care rodent cases, it is advisable to prepare an emergency kit and supply area that is stocked, maintained, and readily accessible. Having designated locations assigned for equipment and kits, containing supplies that are replaced on a periodic basis, will prevent loss of time looking for relevant materials during an emergent situation (Bergdall and Green, 2004, Mader, 2002, Paul-Murphy, 1996).

Items to consider maintaining for rodent procedures and surgery include the following:

- **Gram scale with disposable weigh boats:** For easy collection of body weights of individual rodents to then accurately calculate anesthetic and analgesic doses.
- **Surgical tables:** Design should be easily sanitizable and appropriate for routine rodent surgeries, which can be performed in dedicated areas outside a formal surgical suite; enhanced design features for a table could include adjustable height for the surgeon and allowance for tilting.

- **Heating sources:** Warm water-recirculating heat blankets, warm-air devices like the Bair-Hugger® blanket and ThermoCare® isolators, microwaveable gels, pads, and pockets may also be of use for rodents (please see Chapter 1 for further details on supplemental heat provision). *Electrical heating pads should not be used due to the potential for dermal burns caused by uneven heating.*
- **Anesthesia machines:** Equipment should include a vaporizer, oxygen tank and holder (or some connection to a central O_2 source with flowmeter), and CO_2 absorber. Vaporizers need to be specific to the type of volatile anesthetic used. Machines designed to permit simultaneous anesthesia and monitoring of multiple rodents are available.
- **Anesthesia ventilators:** For procedures like imaging or prolonged surgeries, mechanical ventilation is typically necessary. Ventilators are generally pneumatic (with bellows); however, electronic ventilators are available for rodents.
- **Anesthesia monitoring:** To assist with determining the overall ventilation status of the patient, consider investment in ECG monitors for visual and auditory outputs, pulse oximeters, and capnographs for measurement of the expired CO_2 concentration (Stanford, 2004).
- **Blood pressure monitors:** These provide information about the hemodynamic status of the patient and are especially useful for cardiovascular surgeries.
- **Cautery:** Cautery applies an electric current to cut through tissues and seal vessels as a means of hemostasis; harmonic scalpels may be an alternative, using ultrasound to cut and coagulate tissues simultaneously. For rodents, cordless cautery loop tips (e.g., MediChoice®) are extremely useful for tissue amputations and hemostasis.

euthanasia considerations

Euthanasia is described by the American Veterinary Medical Association (AVMA) as a method of killing that minimizes pain, distress, and anxiety experienced by the animal prior to the loss of consciousness, and causes rapid loss of consciousness followed by cardiac or respiratory arrest and, ultimately, a loss of brain function (AVMA, 2013). Unfortunately, euthanasia is often the elected outcome for many critically ill laboratory rodents due to irreversibility of disease and request for diagnostic necropsy. To alleviate prolonged suffering, interventional euthanasia is always preferable to spontaneous death of laboratory animals. The AVMA provides euthanasia guidance for the entire veterinary profession; however, the laboratory animal community is expected to adhere to this published guidance and additional references (AVMA, 2013, Danneman et al., 2012, Muir et al., 1989, National Research Council, 2011, Sharp and Villano, 2012).

Timeliness of humane endpoints involving euthanasia will need to be outlined for critical cases, in particular if research data are dependent on the preservation of the animal tissues. It is recommended that staff achieve a consensus for when rodents should be euthanized and in what time frame after reporting on the critical health condition (immediate, within 2 h, within 24 h, etc.). The selected euthanasia method should minimize sources of potential distress (excessive handling, disruption of compatible housing groups, etc.). Methods may be deemed as *acceptable* (considered to reliably meet the requirements of euthanasia), as *acceptable with conditions* (considered to reliably meet the requirements of euthanasia when specified conditions are met), or as *unacceptable* (does not meet the requirements of euthanasia) (AVMA, 2013). Assurance that animals are to be euthanized properly must be documented in the animal use protocol approved by the Institutional Animal Care and Use Committee (IACUC).

General considerations for selection of euthanasia method for laboratory rodents include the following:

1. Size, weight, age
2. Need for and type of physical restraint
3. Time required to produce loss of consciousness and death
4. Reliability and irreversibility
5. Skill level of personnel
6. Availability of facilities, equipment, and drugs

Acceptable methods of euthanasia for laboratory rodents include injection of barbiturates (typically intraperitoneally at a dosage of three times what would be used for anesthesia), injection of barbiturates in combination with local anesthetics and anticonvulsants, and injection of lethal doses of dissociative agents (such as ketamine) in combination with drugs like xylazine or diazepam (AVMA, 2013).

Methods that are acceptable with conditions include the use of inhaled anesthetics (isoflurane, sevoflurane, desflurane, with or without nitrous oxide in combination); inhaled CO_2 (with or without inhaled anesthetics; provided at a flow rate to displace 10–30% of the chamber or cage volume per minute); and inhaled carbon monoxide (not commonly used in the laboratory animal setting). Note that euthanasia chambers prefilled with carbon dioxide are *unacceptable*. Noninhaled agents like tribromoethanol and ethanol are acceptable with conditions. Physical methods (e.g., cervical dislocation for animals < 200 g BW, decapitation, and focused beam microwave irradiation) are also acceptable with conditions (AVMA, 2013).

Unacceptable methods include the exposure of conscious mammals to inhaled nitrogen or argon; animals would have to be under heavy sedation or anesthesia for these agents to be allowed, and other euthanasia methods are preferable. Noninhaled agents that are unacceptable as a sole agent of euthanasia would include potassium chloride, neuromuscular blocking agents, opioid overdoses, urethane, or α-chloralose. There may be circumstances when these agents, in combination with other methods, may be permissible (AVMA, 2013).

Commonly, CO_2 overdose is utilized for euthanasia of laboratory mice and rats. Because there may be a possibility that rodents can recover from a deeply anesthetized state following exposure to CO_2, there is a need to ensure a secondary (confirmatory) means of euthanasia. Further guidance is provided by the Office of Laboratory Animal Welfare (OLAW), through the Public Health Service arm of the federal government (see relevant section in Chapter 5 "Resources and Additional Information"). These confirmatory methods, which should be performed after CO_2 overdose, might include one of the following: exsanguination, decapitation, cervical disarticulation, bilateral thoracotomy, or at least 50% additional time in the euthanasia chamber filled with 100% CO_2 (IACUC-UPENN, 2008).

A recommended method for euthanasia of a pregnant dam is CO_2 exposure followed by cervical disarticulation. If fetuses are not required for study, the method chosen for euthanasia of a pregnant female should ensure cerebral anoxia to the fetus and minimally disturb the uterine milieu to minimize fetal arousal (Klaunberg et al., 2004). Euthanasia of neonates requires greatly prolonged exposure to inhalant anesthetics, including CO_2 (Table 4.1). Resistance to hypoxia results in a prolonged time to unconsciousness if inhalant agents are used. Therefore, it is an acceptable method for euthanasia

Table 4.1: Recommended Euthanasia Times for Laboratory Mice and Rats

	Minimum Time in 100% CO_2 (minutes)	
Age	**Mice**	**Rats**
Nonhaired pups: 0–6 days	60 min	40 min
Haired pups, eyes closed: 7–13 days	20 min	20 min
Haired pups, eyes open, preweaning: 14–20 days	10 min	10 min
Weanlings and adults: 21+ days	5 min	10 min

Source: Data from Pritchett-Corning, KR. 2009. *J Am Assoc Lab Anim Sci* 48:23–27; and Pritchett, K, Corrow, D, Stockwell, J, and Smith, A. 2005. *Comp Med* 55:275–281.

of neonates to administer sufficient quantities of injectable chemical anesthetics (American College of Laboratory Animal Medicine [ACLAM], 2005, AVMA, 2013). Special cases for euthanasia are further described in the AVMA guidelines (AVMA, 2013).

experimental autoimmune encephalomyelitis and demyelinating disease model considerations

Rodent models of experimental autoimmune encephalomyelitis (EAE) and related demyelinating diseases may result in a complex spectrum of acute, chronic, and relapsing-remitting disease courses that most often result in varying degrees of progressive ascending paralysis. Due to the extreme variability in the onset and progression of clinical signs and disease course, close monitoring, assessment of body condition score (BCS), and provision of supportive care are necessary for EAE animals (IACUC-UPENN, 2013).

EAE Scoring

Clinical signs and ascending paralysis in EAE are commonly assessed on a six-stage scale of 0–5, with 0 representing a clinically normal condition and 5 representing paralysis of all limbs (quadriplegia). Other scoring systems may be preferable and should be clearly defined in the protocol and made available to animal care staff in close proximity to the animal housing room.

0: Clinically normal

1: Decreased tail tone or weak tail only

2: Hind limb weakness (paraparesis)

3: Hind limb paralysis (paraplegia) or urinary incontinence
4: Weakness of front limbs with paraparesis or paraplegia (quadriparesis) or atonic bladder
5: Paralysis of all limbs (quadriplegia)

Animal Care

Verification that research personnel are properly trained in the procedures related to these disease models must be documented in the IACUC protocol. It is preferable to keep a written record of the disease progression with information including the start date of experiments, the BW and overall condition, and the general appearance per the EAE scoring scale presented in the preceding section. Enrichment with cotton nesting material is not recommended for animals that will develop weakness and paralysis as the fibers may entrap and strangulate weakened limbs/tails. It is recommended that a soft bedding substrate (versus corncob bedding) be utilized to minimize skin trauma secondary to paralysis.

When clinical signs are expected to begin, laboratory staff should monitor mice at least once daily. The following guidance is designed to assist with increasing monitoring and measures of care:

- *Score 1–2*: Separate affected animals to another cage to avoid injury by unaffected animals. Alternatively, house with similarly affected animals (preferred) to maintain social housing environment. If available, provide water bottles with elongated sipper tubes, even if housed in autowatering cages. Provide nutritional (pelleted chow, special diets) and fluid gel supplements on the cage floor and replenish as needed or at least two or three times each week. Animal weights and BCS should be determined at least twice weekly.

  Note: If the BCS is 2 or less or if the animal has lost over 10% of baseline weight, administer sterile fluid (1 ml SC) daily.

- *Score 3–4*: Treatments of daily fluid support and nutritional and gel supplements should continue. Continued weight collection and BCS should be performed twice weekly. Once hind limb paralysis or urinary incontinence is noted, increased monitoring and assistance with bladder expression will be expected. Gentle palpation of the bladder should be done daily to express urine; gently press on the caudal abdomen to assist with urination. If the bladder loses tone (atony), express twice daily. Due to the potential for urine dribbling onto skin, it will be critical to assess the animal for ventral and perineal dermatitis, urine scald, penile prolapse, and tail lesions. If these clinical abnormalities appear, the animal will require further treatment, made in consultation with the veterinary staff, to minimize urine scald.

- *Score 5*: Once an animal becomes quadriplegic (paralyzed in all four limbs), it should be euthanized promptly unless the approved IACUC protocol states otherwise. Other criteria for euthanasia include a loss of more than 20% BW, a BCS of 1, or moribund state.

fasting considerations

Small mammals require an almost-continuous supply of food and water; accordingly, fasting (withholding of food for a designated period prior to testing, then return of food) or restricting (limiting ration of food provided) should be minimized to the extent necessary to achieve the scientific objectives while maintaining animal well-being. It is notable that many research protocols will request a time period for the fasting of laboratory mice prior to procedures; fasting by this definition means that the animals will be allowed free access to water but that food may be removed prior to a planned procedure. Fasting may occur for a number of reasons, including minimizing the variability of drug exposure time prior to necropsy and reducing the contents of the gastrointestinal tract prior to intraperitoneal injection, intragastric dosing, or gastrointestinal surgery. In toxicology laboratories, food may be withheld from rodents prior to necropsy to improve the ease of handling and fixation of the gastrointestinal tract and to yield more uniform liver histology sections. Withholding of food is nonphysiological and may compromise ease of collection of biological samples and overall animal condition. This practice also may contribute unnecessary stress to experimental animals (Turner et al., 2001).

In general, presurgical (~16 h or longer, also known as "overnight") fasting is *not recommended* for mice and rats, particularly due to the increase in food generally ingested by rodents at the beginning of the dark cycle. As well, water *should not* be withheld prior to anesthesia (Lester et al., 2012). Rodents do not vomit; therefore, the rationale for fasting to prevent aspiration pneumonia is not pertinent to mice and rats. As well, the high metabolic rate can lead to hypoglycemia and liver changes if rodents are fasted for any length of time (Morrisey, 2003a). Certain institutions may have policies that prevent any fasting of animals prior to surgery unless the feeding condition is a key aspect to the experimental model (Toth and Gardiner, 2000). *In general, it is recommended that mice and rats be fasted for no more than 2 to 4 h prior to procedures that require an emptied stomach* (Lester et al., 2012).

The provision of a palatable, simple carbohydrate to rats overnight, in the form of sucrose (sugar cubes), reduces the size of the gastrointestinal tract while minimizing other side effects of food withholding, such as alterations in serum biochemistry parameters and body weights (Levine and Saltzman, 1998). Offering sugar cubes represents an inexpensive, simple, and readily available alternative to overnight fasting. However, the overnight feeding of sucrose, in lieu of limiting chow or complete fasting, can result in marked changes in gastrointestinal tract weight and pancreatic and hepatic structure and function, as described for laboratory rats (Turner et al., 2001).

The experimental rationale for fasting of rodents is typically related to behavioral motivation and assessments. Animals may experience some discomfort during longer fasting periods, and the IACUC would require scientific justification for a particular duration of deprivation balanced against the induction of potential distress or physiologic harm (Rowland, 2007). Restriction studies normally are preformed on healthy animals; thus, the physiologic consequences differ from those of anorexia caused by illness. A healthy animal that has lost 15% of body weight by restriction is likely to acclimate and become clinically stable, whereas one that has lost the same weight due to illness is typically not stable (Rowland, 2007). Overall, rodents can acclimate to fasting for experimental purposes, specifically by efficiently reducing further fluid or energy losses through a combination of innate behavioral and physiologic adjustments.

fluid therapy considerations

Administration of warmed (~37°C) fluids is a frontline intervention to treat acutely or chronically ill animals with various diseases, including electrolyte disorders, acid-base disorders, and hypovolemia induced by blood loss. Administering subcutaneous fluids in critically ill rodents is the most common, efficacious, and least-invasive method of delivery (DiBartola, 2000, Hawkins and Graham, 2007, Klaphake, 2006, Mader, 2002). Dehydration (a loss of body fluid from intracellular, plasma, interstitial, or transcellular compartments) is the key indicator for fluid therapy. Dehydration may be noted in animals that are in poor body condition, those that have disease, or those that have undergone prolonged surgery performed under anesthesia. Prompt and appropriate fluid therapy should be instituted prior to or at the time that dehydration is documented.

Fluid requirements for dehydration (Hawkins and Graham, 2007) are calculated as follows:

$$\% \text{ Dehydration} \times \text{kg} \times 1000 \text{ ml/L} = \text{Fluid deficit (L)}$$

Published dosages for fluid therapy in rodents range from 30 to 90 ml/kg in the first hour; further, maintenance rates for critically ill rodents have been calculated at 100 ml/kg/day SC or IP with compensation for special losses (Mader, 2002, Paul-Murphy, 1996).

Crystalloids

Crystalloids are most often used for rodent fluid therapy and are defined as isotonic solutions (with plasma) that contain both electrolytes and nonelectrolytes and are capable of entering all of the body fluid compartments. Crystalloids are equally as effective at increasing blood volume as colloid fluids but must be administered in greater amounts since they are absorbed within all fluid compartments. Crystalloids can be classified as replacement fluids (if they are similar to the extracellular fluid) or as maintenance solutions (if they contain less sodium and more potassium).

- *Ringer's solution with (LRS) or without lactate.* Ringer's solution contains sodium, chloride, potassium, and calcium. LRS is useful in increasing tissue perfusion and in extracellular blood expansion. In addition, the lactate counteracts metabolic acidosis that may occur with kidney failure or acute fluid loss. There may be unwanted complications if LRS, which contains calcium, is administered concurrently with other drugs (Hackett and Lehman, 2005).
- *Normosol®-R.* Normosol-R is a buffered solution similar to LRS; it contains acetate and gluconate instead of lactate.

- *Saline.* Normal saline (0.9% NaCl) is an isotonic solution that contains normal sodium concentrations but greater chloride concentrations than body fluids. Saline is used for acute extracellular volume expansion.
- *Dextrose.* Dextrose at 5% (diluted in either NaCl or LRS) provides about 200 kcal/L and may satisfy the maintenance energy requirement for rodents.

Colloids

Colloids are much less commonly used in laboratory rodents and are defined as fluids that contain large macromolecules that are restricted to the plasma compartment and cannot enter any of the body's fluid compartments. Colloids should be used in patients with shock and hypoalbuminemia, particularly if an intravenous or intraosseous access route has been established, to achieve volume expansion rapidly. Colloids include natural (e.g., plasma or whole blood) or synthetic (dextran, hydroxyethyl starch [Hetastarch], and stroma-free hemoglobin [Oxyglobin®]) formulations. There are noted limitations with colloid fluid therapy, including anticoagulation activity; little published evidence of their effectiveness in rodent patients exists.

food and fluid regulation procedures

Food and fluid regulation (whether scheduling or restriction) for rodents is typically used in the research setting for three main areas of study: as a means to motivate animals to perform novel or learned tasks; to analyze the motivated behaviors and physiologic mediators of hunger and thirst; and to investigate homeostatic regulation of energy metabolism or food balance. Scientific justification should be provided for using food and fluid regulation, and the least regulation necessary to achieve the scientific objective should be used.

Experiments that involve food or fluid regulation should evaluate the following factors: the level of regulation (meaning how limited the access will be to food and fluid), potential adverse consequences of regulation, and methods for assessing the health and well-being of animals undergoing regulation (National Research Council, 2011). Investigators should provide to the IACUC plans for the implementation of sufficient and scheduled monitoring (routine weighing and target weight) of animals during food and fluid regulation studies.

Rodents should be acclimated over time (a minimum of 3 days) to food and fluid regulation paradigms. Consideration should be made to allow food and water to be available concurrently as rodents typically do not eat caloric requirements without available water. Regulated levels of food should not be lower than 30% of ad libitum values. Overall, experiments involving food and fluid regulation are not recommended in rodents less than 14 weeks of age.

At times, the degree of restriction may actually be better described as deprivation for up to several hours. Studies have shown that fluid deprivation (with ad libitum food access) in mice for 12, 24, and 48 h results in average weight losses of 9%, 12%, and 18% of initial body weight, respectively. These animals (relative to nondeprived controls) have decreases in plasma volume and increases in plasma renin activity and corticosterone. For mice chronically restricted over the course of 7 days to a 50–75% water ration, modest dehydration anorexia (food intake reduction of ~10%) has been noted, with severe renal lesions identified at necropsy. Acute restriction of fluid over 24 h is associated with significant physiologic stress and is not recommended due to welfare impact (Bekkevold et al., 2012).

Animals on regulation should be closely monitored and weighed at least weekly and weighed more often if animals are undergoing greater restrictions (National Research Council, 2011). Weight and BCS should be compared to age- and strain-matched control animals.

For animals undergoing scheduling or regulation, written records should be maintained daily to document food and fluid consumption, hydration status, and any behavioral or clinical changes used as criteria for temporary or permanent removal of an animal from a protocol (National Research Council, 2011). It will be useful to have these written records readily available for the animal care staff, IACUC members, or any outside reviewers. Designated personnel should document certain experimental and clinical information, such as the following:

- Date (daily documentation is necessary for animals undergoing regulation)
- Baseline weight (prior to initiation of restriction)
- Weight and BCS (twice weekly)
- Indication of schedule
- Indication that access to food and water was granted (daily)
- Overall health, behavior, and activity (daily)

Specific humane, experimental, and interventional endpoints must be clearly stated in the IACUC protocol. For food regulation in rodents, the animal should not lose more than 20% of control weight or baseline BW (if adult) matched by age, strain, and sex unless scientifically justified to do so. Veterinary staff should be involved in evaluations of animals that have lost 20% or more of baseline weight. As well, it is not recommended that a food- or fluid-regulated animal have a BCS of 2 or less if the BCS was higher than this level at the start of the regulation protocol. For fluid regulation, animals with a weight loss of 10% from baseline weight should be considered clinically dehydrated and be treated as outlined in the discussion that follows. Any rodent appearing dehydrated (displaying listlessness or inactivity, with an increased "skin tent" time) should have a measured volume of fluid provided promptly; care should be taken not to fluid overload an animal that has been acclimated to restriction. In addition, up to 2 ml of fluid can be administered subcutaneously to boost hydration status and improve animal well-being. Research personnel involved in water restriction studies should be trained appropriately to identify dehydration and correctly administer subcutaneous fluids.

Aspects of this section were adapted from institutional documents (IACUC-UPENN, 2011a).

humane or "clinical" endpoint considerations

According to the National Research Council guide, experiments that may result in "severe or chronic pain or significant alterations in the animals' ability to maintain normal physiology, or adequately respond to stressors, should include descriptions of appropriate humane endpoints" (National Research Council, 2011, p. 5). Humane interventions or endpoints are defined as the point at which pain or distress in an experimental animal is prevented (with therapeutic interventions), terminated (by cessation of participation in the study), or relieved (typically by euthanasia). Selection of appropriate interventions provides significant opportunities for refinements; as well, these endpoints must be relevant to the selected model. For example, tumorigenicity studies could be terminated as soon as progressive tumor growth is documented; however, carcinogen-induced papillomas may require later endpoints so that cells are able to transform into a malignant state (Workman et al., 2010).

It is extremely useful to itemize and describe specific humane endpoints in IACUC protocols, particularly for those procedures that may involve potential pain and distress. Humane interventions and endpoints should be determined by a collaborative effort of research staff with laboratory animal veterinarians. Humane endpoints for a variety of models are further reviewed in an issue of the *Institute for Laboratory Animal Research Journal* ("Humane Endpoints," 2000).

Studies that commonly require special consideration for endpoints may include the following:

- Tumor development
- Infectious disease
- Vaccine challenge
- Pain and trauma modeling
- Monoclonal antibody production
- Assessment of toxicological effects
- Organ or systemic failure
- Models of cardiovascular shock
- Demyelinating diseases
- Generation of animals with abnormal phenotypes

To develop a humane endpoint, one should be aware of the clinical progression that a particular (group of) animal(s) is likely to experience as a result of experimental manipulation or any spontaneously occurring disease that might develop during the lifetime of the animal(s). Personnel must be adequately trained in the basic principles of laboratory animal science (National Research Council, 2011); in this case, staff need to be able to recognize species-specific behaviors, as well as signs of pain, distress, and morbidity.

The selection of appropriate humane endpoints requires a detailed knowledge of the impact of the procedure on the animal to allow for intervention before unpredicted distress or pain develops. If the outcome of a particular experimental model is not known, which may be the case if studies are novel or alternative endpoint information is lacking, pilot studies can be an effective method for identifying and defining humane endpoints and reaching consensus among the investigators, the IACUC, and veterinary staff (National Research Council, 2011). Pilot studies can be designed to assess both the procedure's effects on the animal and the skills of the research team and must be conducted with IACUC approval.

Various clinical signs are indicative of a moribund condition in laboratory animals. If any of these signs are noted, prompt consultation with veterinary staff or euthanasia should occur (Aldred et al., 2002, IACUC-UPENN, 2011b, Madeddu et al., 2006, Nemzek et al., 2004, Paster et al., 2009, Schenk et al., 2012):

- Any condition interfering with eating or drinking (e.g., difficulty with ambulation)
- Rapid weight loss or net weight loss of more than 20% of BW
- Prolonged inappetance
- Evidence of muscle atrophy
- Marked loss of body condition
- Diarrhea, if debilitating, or constipation
- Markedly discolored urine, polyuria, or anuria
- Roughened hair coat, hunched posture, lethargy, or persistent recumbency
- Central nervous system disturbance: head tilt, seizures, tremors, circling, paresis, paralysis
- Lack of mental alertness
- Coughing, labored breathing, nasal discharge, or respiratory distress
- Jaundice or anemia (pallor)
- Bleeding from any site
- Excessive or prolonged hyperthermia or hypothermia
- Marked dehydration
- Measurable distention of abdomen or torso

Frequency of animal assessments and monitoring, as well as the objective criteria to be used for health evaluations, should be clearly described in any IACUC protocol. Collection of individual assessments (scoring) on fur appearance, respiration rate and effort, mobility and ambulation, behavior, and body condition will all contribute to the overall decisions about enacting humane interventions and endpoints for animals.

nutritional therapy considerations

Anorexia, while not always observed, typically manifests as a loss of body condition and weight in critically ill rodents. Nutritional support is key in the care of rodents, as the healthy rodent requires a maintenance rate of 150–350 kcal/kg/day. The nutritional requirement can be two to three times this maintenance rate when animals are ill (Paul-Murphy, 1996).

Nutritional supplementation of high-quality, laboratory-certified feedstuffs, when provided to debilitated rodents, is part of expected supportive care. Modifications of the animal's own diet (typically rodent block chow) can be done with the addition of Nutri-Cal® (a high-calorie, palatable veterinary gel supplement) or soaking/softening the chow in Ensure® (a high-calorie, flavored human shake supplement). For animals in poor health, food and fluid options are often provided on the cage floor in disposable trays/dishes or can be delivered in liquid format by oral gavage for direct instillation into the stomach. Oral gavage of mice (Figure 4.2) using sucrose-coated needles has been validated as a means to decrease stress and improve animal welfare during the gavage process (Hoggatt et al., 2010).

An expansive selection of nutritional and fluid supplementation products is available for laboratory rodents that are in suboptimal health. Syringe feeding will likely be tolerated if the animal is not moribund. Nasogastric intubation is typically physiologically impossible in mice and may have limited applications in the critically ill rat. The cachectic or critical patient may require more calories, a more concentrated caloric diet, and high palatability until the appetite is restored and the animal is eating on its own (Klaphake, 2006). Supportive and routine feedings may be necessary at high frequency until rodents are through the crisis phase of their illness.

Specialty nutritional supplementation (e.g., Nutra-Gel™ Diet, Bio-Serv®, Frenchtown, NJ, http://www.bio-serv.com/) can be provided to boost palatability and caloric intake for rodents that are ill or have undergone major surgery or for models with abnormal phenotypes that affect mastication (Figure 4.3).

Independent studies conducted by the particular vendor companies have ascertained that rodents find flavors like bacon and molasses particularly appealing for these products (e.g., Transgenic Dough Diet™, Bacon Softies™, Supreme Mini-Treats™, Bio-Serv), and they may also be used effectively in pregnant females (Love Mash™

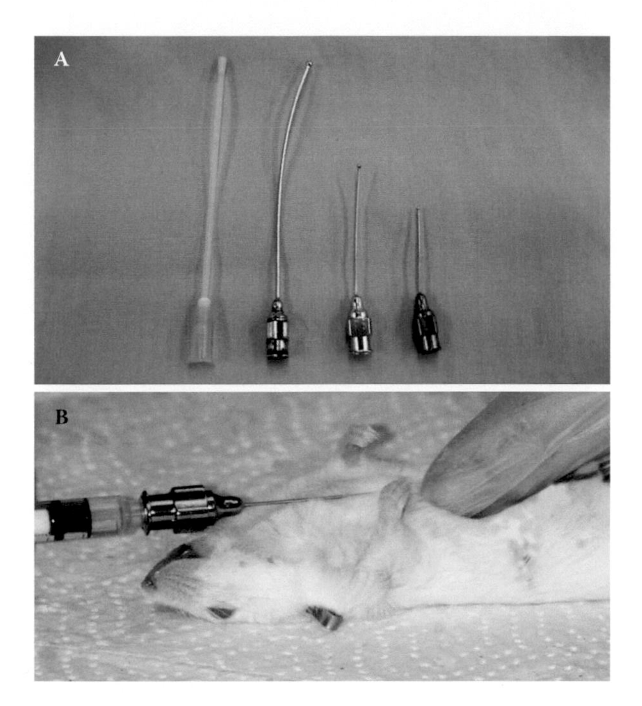

Fig. 4.2 Oral gavage in mice. (A) A variety of plastic and stainless steel gavage needles is available for dosing in rodents and can be straight or curved. Flexible needles (left) reduce the likelihood of inadvertent esophageal rupture but can be chewed by the animal. Ball-tipped, curved (rat sized, middle left) and straight (mouse-size, middle right and right) needles are readily sanitized but can induce injury if their passage is forced. (B) Measuring the gavage needle for appropriate length for dosing, from tip of nose to last rib. The needle can be marked for easy visualization of the appropriate insertion distance. (Reprinted with permission from AALAS. Turner, PV, Pekow, C, Vasbinder, MA, and Brabb, T. 2011. *J Am Assoc Lab Anim Sci* 50:614–627.)

Rodent Reproductive Diet, Bio-Serv) and for administration to dams to prevent cannibalism of litters (Figure 4.4).

Gel supplements can provide palatable choices to improve postsurgical weight gain and assist with boosting calories as well as hydration and can be formulated to assist with oral analgesic delivery (e.g., DietGel® Recovery, DietGel® Boost, MediGel® Sucralose, ClearH_2O®, Portland, ME, http://clearh2o.com/) (Figure 4.5). If animals require further or continuous fluid supplementation in addition to those provided as a bolus subcutaneously or intraperitoneally, oral options

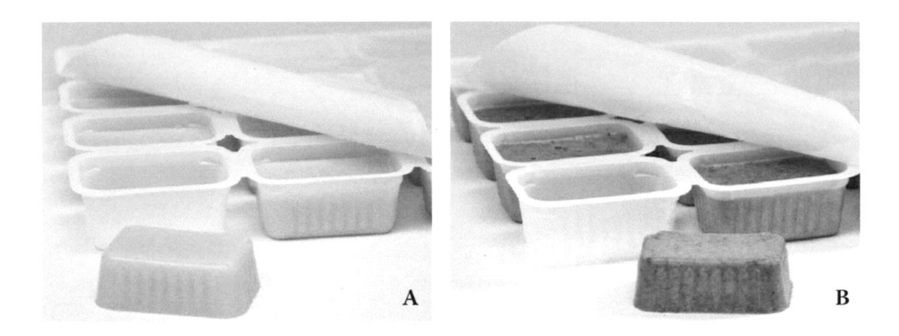

Fig. 4.3 Supplementary nutritional products, including (A) NutraGel™ Diet purified and (B) grain-based options. (Photos courtesy of Bio-Serv®.)

Fig. 4.4 Supplementary nutritional products, including (A) chocolate-flavored Supreme Mini-Treats™, (B) Bacon Softies™, (C) Love Mash™ Rodent Reproductive Diet, and (D) Transgenic Dough Diet™. (Photos courtesy of Bio-Serv®.)

can include water provided in individual packages inside the cage (HydroPac®, Seaford, DE, http://www.hydropac.net/) or as gel formulations (e.g., HydroGel™, ClearH_2O®; and Napa Nector™, Systems Engineering, Napa, CA. http://www.selabgroup.com/welcome.htm).

Providing pelleted food in powdered form to young C57BL/6 mice has been shown to significantly affect ingestion; animals consumed more food when presented in powdered form than when it was presented as pellets. The significant difference was associated

Fig. 4.5 Fluid and diet supplements, including DietGel® Recovery, HydroGel®, and DietGel® 76A. (Reprinted with permission from ClearH_2O®.)

with corresponding higher intake of nutrients, including calories, from the powdered forms. Therefore, it is key to also consider the physical form of the diet when providing supplementary nutrition to minimize confounding influences and experimental variations (Yan et al., 2011).

perioperative care considerations

The "Guide for the Care and Use of Laboratory Animals" emphasizes that successful surgical outcomes require appropriate attention to presurgical planning, personnel training, anesthesia, aseptic and surgical techniques, assessment of animal well-being, appropriate use of analgesics, and monitoring of the patient's physiologic status during the procedures and beyond. There is an expectation that training aspects of perioperative care will be provided to inform research personnel performing surgery about maintenance of aseptic technique during survival surgery (National Research Council, 2011). A dedicated rodent surgical facility is not required, yet a designated animal procedure area is necessary and should be located where it can be maintained in an orderly, noncluttered status at all times during surgery to minimize potential contamination of the patient. This dedicated area should be away from heavy personnel traffic flow and other unrelated activities and should have a surgical surface constructed of a readily disinfected material.

Aseptic surgery is performed with sterile gloves and sterilized (autoclaved) instruments and materials (Figure 4.6) and takes precaution

Fig. 4.6 Perioperative considerations include provision of a sterile pack with an enhanced draped space to allow for placement of instruments and a larger sterile working surface. (Images courtesy of University of Pennsylvania; S. Volk.)

to avoid introduction of infectious microorganisms to the patient. Assurance that research personnel are properly trained in surgical preparation and technique must be documented in the IACUC animal use protocol; personnel should wear sterile gloves and masks while performing surgery. Gloves must be replaced if aseptic technique is disrupted (e.g., touching the isoflurane vaporizer with sterile gloves, moving the animal with sterile gloves). With proper planning, simple survival rodent surgeries may be performed by one person. If this cannot be accomplished because of the complexity of the procedure, then to consistently maintain aseptic technique, there should be a surgical assistant or anesthetist who is trained to perform such tasks that would otherwise interfere with proper aseptic technique. If it is necessary for the surgeon to leave the surgical area during a procedure, then they must reglove before resuming surgery.

Ensuring surgical procedures are kept sterile will lead to a decreased chance of subclinical infections postoperatively or other adverse postoperative outcomes. In addition, tissue should be handled gently, with appropriate size instruments for rodent surgery.

Rodent preparation prior to surgery typically includes the removal of the hair coat in a wide area around the intended incision site. Excess clipped or depilated hair and other gross debris can be wiped from the area using sterile gauze or a small disposable alcohol pad;

however, dousing the animal with alcohol is *not recommended* due to potential for hypothermia as the alcohol evaporates from the skin. The area can then be disinfected with appropriate surgical scrub. Alcohol alone is not an appropriate disinfectant. Iodophors (e.g., betadine) or chlorohexidines may be used and then wiped away with sterile warmed or room temperature saline. Preemptive analgesia will help to suppress pain responses that may be experienced postoperatively, particularly through administration of opioids prior to initiation of surgery (Flecknell, 2001). As well, and as mentioned throughout the text, ocular protection with artificial tears (Rugby® Sterile Artificial Tears Ointment Lubricant–Ophthalmic Ointment), vitamin A ophthalmic ointment, or sterile lubricant (e.g., PuralubeTM) is recommended for any animal undergoing anesthesia (Figure 4.7).

Attention to appropriate draping can increase the potential to maintain a sterile surgical field. It is recommended that transparent surgical drapes be utilized to ensure the patients can be observed and monitored under anesthesia (Figure 4.8). As well, adhesive drapes are being used increasingly with rodent patients to help affix the anesthetic-delivering nose cone to the animal and the patient to the surgical surface.

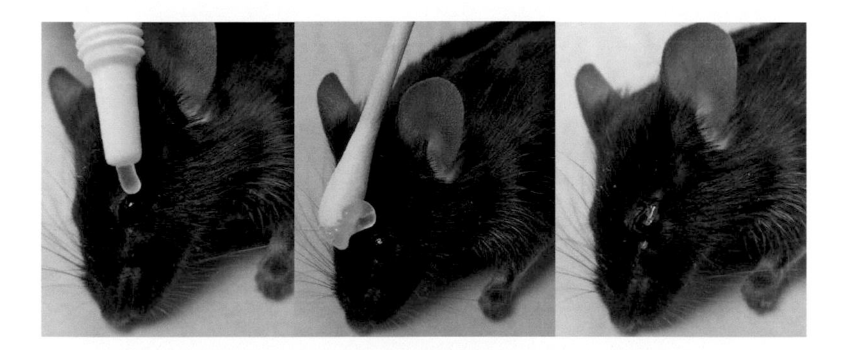

Fig. 4.7 Protective ophthalmic ointment should always be applied to mice and rats undergoing anesthesia for any length of nonsurgical or surgical procedure. Once the animal is sedated, ointment (Rugby® Sterile Artificial Tears Ointment Lubricant–Ophthalmic Ointment) can be dripped directly onto the eye from the application vial (left) or transferred to a sterile swab and then touched to the eyes (middle). The eyes typically remain open (right) during anesthesia, and the ophthalmic film will provide a protective layer to prevent desiccation of the globe and surrounding structures. (Images courtesy of the University of Pennsylvania, ULAR.)

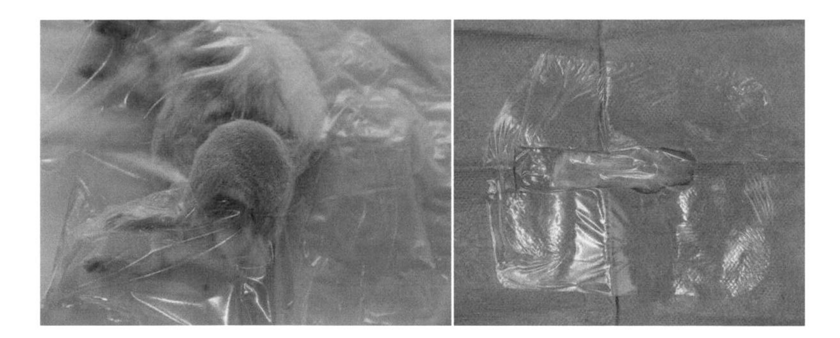

Fig. 4.8 A rodent leg draped with transparent adhesive Tegaderm™ dressing to permit an enhanced view of the surgical site (left); the animal can be draped first with a standard disposable blue cloth and the adhesive used to hold the drape in place (right) while permitting a larger sterile work space for the surgeon. (Images courtesy of University of Pennsylvania; S. Volk.)

Advantages of this level of restraint include insurance that the animal will not bccomc disconncctcd from thc sourcc of ancsthcsia and that there will not be mobility of the patient during surgical manipulations (Locke et al., 2011). Adhesive drapes (i.e., Tegaderm™ Transparent Dressing, 3M, Minneapolis, MN) can be ordered in varying sizes as they are typically applied to human wounds. These sterile film dressings provide permeability for oxygen and moisture exchange while limiting access of debris and pathogens to the incision site. For mice, drape sizes of 4 × 4¾ inches are appropriate for an abdominal procedure; rats will typically require a larger adhesive drape size of 6 × 8 inches. Anecdotally, sterile over-the-counter plastic wrap (e.g., Saran™ wrap) has been used successfully as a transparent rodent drape. For the novice surgeon, a larger sterile drape can be placed first, with the adhesive transparent drape placed over the surgical window to hold the larger drape in place, which provides a wider area of sterility for instrument placement and suture storage, yet the animal can still be visually monitored for respiration and movement during the procedure. Due to the oxygen-rich environment of the surgical area, coupled with the potential to use cautery or other electrical devices within close proximity, one should practice fire safety measures to avoid hazards that could harm personnel and animals (Caro et al., 2011, Klein, 2008).

Surgical monitoring includes confirmation of anesthetic depth to an appropriate level for the procedure. For most rodents, a brief and strong manual pressure on the toes ("toe pinch") should not elicit a withdrawal response if the animal is at a surgical plane of anesthesia. There should also be no palpebral response, confirmed by gently tapping the medial aspect of the rodent eyelid without eliciting a blink response. If anesthesia becomes light (and the rodent potentially conscious) during a surgical procedure, the animal may move in response to skin incisions and tissue manipulations and exhibit a rise in heart and respiratory rates. Further attempts to complete the procedure must cease until the animal is returned to a surgical plane of anesthesia. Anesthetized animals should never be left alone during the surgical procedure. Tissue manipulations should be performed gently to reduce the degree of postoperative pain; stretching and pulling on sensitive structures triggers nociceptive nerve endings, and these impulses may heighten pain levels following recovery from anesthesia (Flecknell, 2001).

Postsurgical care management should include collaborative assistance between veterinary and research staff members. Sutures are more likely to remain intact, without chance of evisceration, if absorbable sutures or skin staples are used. Incisional hernia is the most common abdominal wall defect, usually formed by trauma or infection. Defects can be patched with synthetic materials, such as polypropylene mesh (Sepramesh™) inserted in a "sandwich" form between injured peritoneum and abdominal muscles. The mesh can be combined with Seprafilm®, a bioresorbable membrane that adheres and promotes the normal healing process. Adhesions and inflammation have been documented to be reduced when these membrane products were used for abdominal wall repairs (Esfandiari and Nowrouzian, 2006).

Animals should be placed into a clean cage space, notably without other awake animals, until fully recovered from anesthesia. Housing animals individually after surgery may be necessary to promote anesthetic recovery and to prevent potential for injury from cage mates; however, both the physical and the social environment may affect the way in which an animal copes with the stress associated with postoperative recovery.

The recovery cage can be placed on top of a heat source, or heat supplementation can be provided until full recovery from anesthesia is observed. Fluid gel and soft food (standard food pellets soaked in sucrose water) for up to 4 days postoperatively can decrease potential for dehydration and extreme weight loss; supplemental subcutaneous fluids may also be useful to support postprocedural recovery. Observations and notations of all monitoring and interventions are recommended to occur twice daily for at least 72 h following completion of surgery and recovery from anesthesia (Kalishman et al., 2004). During the postsurgical period, animals should be monitored for signs of pain or distress, as described previously (see Chapter 1). In most species, signs of pain include decreased activity, abnormal posture, increased attention to surgical site, and gait abnormalities. Analgesia and pain relief regimens should be based on the species, the type of procedure performed, the pharmacokinetics of available agents, and any known adverse effects of the specific drugs (Lester et al., 2012). Further information is available in the rodent formulary in Appendix C, along with the additional insights gleaned from published analgesic regimens.

Maintenance of a surgical record for each patient will be helpful to track patient health relative to the presurgical baseline data. Postsurgical documentation (medical records) should include the protocol number, animal identification, observations, date of observation, comments on the general condition and health of animal, and analgesics or other medications given. The specific date, time, and dose of the administered analgesics should be written into the postoperative record. As is expected for a medical record, the surgical record should also include initials of individuals writing the entries.

regulatory considerations

Currently, laboratory mice (of the genus *Mus*) and rats (of the genus *Rattus*) used in biomedical research are exempt from oversight by the Animal Welfare Act and Regulations. However, aspects of their use, care, and treatment are covered in both the *Guide for the Care and Use of Laboratory Animals* (National Research Council, 2011) and the *Public Health Service Policy on Humane Care and Use of Laboratory Animals*, which is specific to coverage of animals used in research funded by the Public Health Service through the National Institutes of Health in the United States (Public Health Service, 1996).

Descriptions of animal models, group sizes, experimental timelines, outcomes, adverse events, and humane endpoints are requested in typical IACUC protocol templates and must be approved by the IACUC *prior* to initiation of any experiments. As well, descriptions of drug types and dosages are required to adhere to veterinary standards of practice. The selection of appropriate sedatives, analgesics, and anesthetics is a moral imperative to minimize, if not eliminate, animal sensation of pain or distress (National Research Council, 2011). Finally, all drugs must be "in date," meaning not used past the expiration date stamped on the vial or package, to adhere with manufacturer recommendations and federal regulations and guidance.

Clinical treatments selected in response to an emergency or critical care situation may not always be included in approved protocols, yet consultation with veterinary staff will permit the limited application of off-protocol drugs and therapeutics for the benefit of overall animal welfare. However, once a clinical treatment has been applied, and if it will be used for additional animals in an experimental cohort, this relevant information *must be amended* into the existing animal care protocol for the IACUC to review and approve.

restraint collar considerations

Collars can be used for prevention of undesirable behaviors (Brown, 2006a, 2006b). Restraint or neck safety collars, most commonly referred to as "Elizabethan collars," have been applied in numerous animal species to prevent coprophagy, self-grooming, licking test compounds off skin, and access to and potential self-injury of postsurgical sites (Figure 4.9).

Considerations in the use of restraint collars include the following:

- Stress induction with placement of the collar, which can result in transient depression in eating and subsequent weight loss or potential for hyperthermia in small rodents.

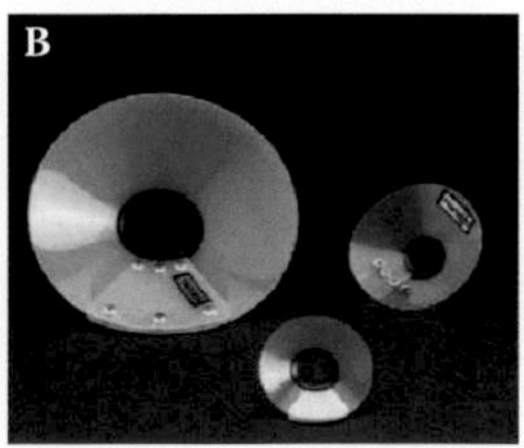

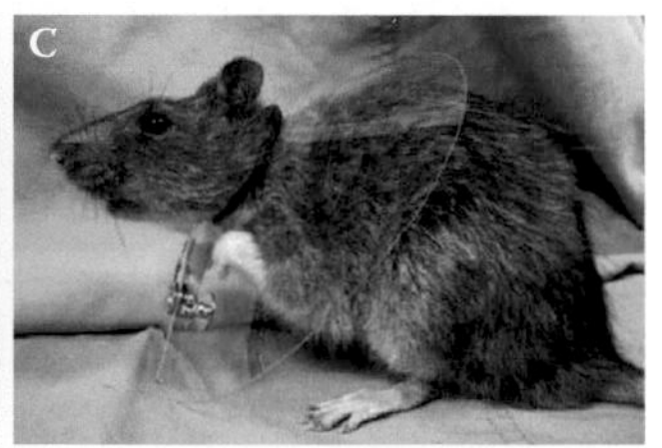

Fig. 4.9 Restraint collars for rodents: (A) low-density polyethylene Elizabethan collars; notice how the neck openings are either bound with cotton jersey for light padding or padded with soft vinyl foam (B). A rat with a caudally directed e-collar (C). Transparent collars that allow an animal to see through them allow for better adjustment than if the collar is made of a solid material or color that obstructs their view. Rodents also perform coprophagy and will need access to consume fecal pellets on the floor of their caging. On a nutritionally complete diet, rats will eat about 10% of their feces. (Reprinted by permission from Macmillan Publishers Limited. Brown, C. 2006a. *Lab Anim (NY)* 35:23–25; and Brown, C. 2006b. *Lab Anim (NY)* 35:25–27.)

- Acclimation period, with increasing time wearing the collar prior to its actual experimental use in the animal, monitored by an observer.
- Fit must be appropriate: not too large to avoid entrapment or further injury and not too small so that the animal chokes or that the skin becomes abraded and ulcerated. *Rule*: Any Elizabethan collar should be able to rotate 360° with minimal difficulty.
- No sharp edges should be left on the collar; attempts should be made to pad the inside edge thoroughly or place a thick layer of tape so the edges are blunted.
- Consider individual housing of the animal during the time when the animal is wearing the collar; there may be intraspecies aggression from the noncollared animals toward the one with the collar.
- It is critical that the animals be able to access food and water sources in their housing/holding areas. Changes to conformation of food and water receptacles may need to be undertaken so that animals can manipulate their faces toward the sustenance without removing the collar.
- Application of aversive nontoxic compounds (metronidazole 500 mg powder mixed with 1 ml New Skin® liquid bandage) to susceptible body parts, in particular following spinal cord injury, has assisted with maintaining animal welfare through prevention of self-injury and autophagia (Zhang et al., 2001).

tracheostomy considerations

As a final or "salvage" procedure for a moribund yet invaluable rodent model, one may attempt to correct respiratory arrest and loss of consciousness by inserting a tracheostomy tube. Due to the critical state of the moribund patient and need to intervene immediately, the tube can be placed under local (topical) anesthesia even in a conscious animal. One can use individual sterile alcohol swabs to part the hair over the ventral neck and visualize the trachea through the skin. A skin incision, using sterile instruments or scalpel blade, should be carefully and gently made just over the area of the trachea, immediately distal to the throat (larynx). Accessing the trachea is best done by gently retracting it into the incision and placing a stay suture underneath; this will help to keep the trachea everted prior to incising between the cartilaginous rings. Once the trachea is isolated, proceed to exert minimal force and gently guide a sterile 16-gauge catheter or 1.0- to 2.0-mm endotracheal tube (if rodent > 100 g) into the distal trachea. Oxygen should then be provided and the animal closely monitored for a return to respiratory function (Paul-Murphy, 1996). The prognosis for this type of procedure is likely poor, given the impaired condition of the patient and any additional physiologic stress experienced by way of this invasive event.

tumor development and monitoring considerations

For those experiments involving tumors, overall tumor burden should be limited to the minimum required to meet the study objectives. The general health and overall condition of the animal is to be assessed with increasing frequency as expected tumors develop. Adverse effects on the rodent will depend on the biology, site, mode of growth of the tumor, and any additional procedures or treatments (Workman et al., 2010). If tumors grow unexpectedly and are not directly related to proposed experiments, research staff should consult with a veterinarian to determine the best course of treatment and interventions. It may be possible to have the tumor excised or treated in some other manner to continue maintaining the animal in a study. Awareness of the types of cancer that may cause inherent pain to the animal should be considered with respect to humane interventions and endpoints (Pacharinsak and Beitz, 2008).

Tumor implantation sites should be chosen to minimize damage to adjacent normal structures; in particular, implantation of tumors on the dorsum or flank of the rodent will likely have limited site-related morbidity. Implantation of tumor cells to the face, limbs, or perineum should be avoided as there is little to no space for tumor growth and expansion. Intramuscular implantation should be avoided as this may be painful due to the distension of the muscle by the tumor. Tumor implantation on the ventral surface of the body should also be avoided due to the risk of irritation to the tumor site in contact with the bedding and floor of the cage. Tumor development should not interfere with normal gait or postures; should not interfere with vital functions (eating, drinking, breathing); should not result in painful responses when palpated; and should not lead to persistent self-trauma (likely secondary to pain). Determining the tumor burden of internal cancers, lymphoreticular tumors, and metastatic disease will be challenging and may require pilot studies in a smaller number of animals to better characterize disease patterns (Workman et al., 2010).

Evaluating Tumor Growth

- Animals that are in a tumor production study should be monitored by the laboratory at least once weekly during the time when the tumor is not yet detectable to observe when tumor growth begins. After a visual or palpable tumor is evident, animals should be monitored by the laboratory at least twice weekly. More frequent observations may be necessary based on tumor growth rate, study parameters, and general condition of the animal.
- Evaluating tumors involving nonsurface areas of the rodent (e.g., bone, brain, lungs, internal organs) can be challenging. Objective tumor size cannot be routinely assessed, and a limited tumor burden (well below the recommended maximum size) may cause impairment and other clinical signs. For tumor models studying non-surface-area tumors, BCS and clinical evaluations of the animals take priority over the measured size of the tumor (Paster et al., 2009). The expected clinical signs and the humane endpoints of those signs must be clearly described in the protocol.
- Evaluating tumor growth on surface area, on the basis of a percentage of BW, is generally inaccurate. While the growing tumor likely will cause an increase in body weight, the general condition of the rodent may be decreased (losing lean body mass), resulting in a relatively stable body weight but a progressively more unhealthy animal. Therefore, tumor growth should be monitored in the context of the evolving BCS, objective dimensional criteria (size) of the tumor burden measured by caliper or other mechanism, tumor location, number of tumors, and tumor ulceration.

The guidance that follows assumes that a normal size adult rodent will be studied (approximately a 25-g mouse or a 250-g rat). The allowable sizes of tumors will be decreased if the tumors are injected into immature or genetically small mice. When on the dorsum or flank of an adult rodent, tumors may be allowed to grow as large as diameters of about 2.0 cm (or 4.2 cm^3) in mice and about 4.0 cm (or 33.5 cm^3) in rats as long as the rodent remains otherwise healthy. The main concerns for permitting individual tumors to develop beyond these recommended size limitations are related to the potential for ulceration of the tumor, with central necrosis of the skin overlying the mass, and potential further injury and poor health.

Tumor Ulceration

Ulceration of a tumor does not necessarily correlate with tumor size or require euthanasia of the animal, but it does typically require more frequent monitoring and treatment of the ulcerated site. Ulceration can lead to discomfort related to the loss of skin integrity or localized infection; as well, hemorrhage at the site of ulceration may occur, and the site may become prone to infection (Narver, 2013). The level of follow-up care for ulcerated tumors is based on both the size of the ulceration and clinical judgment by the veterinarian. Recommendations for monitoring of ulcerated tumors include the following:

- **Pinpoint (< 1 mm) ulcerations** at the site of tumor injection should be monitored at least two times per week for worsening of the ulceration site.
- **Ulcerations (> 1 mm)** of the surface area of the tumor should be monitored at least three times per week and should be reported to veterinary staff for evaluation and potential treatment.

Multiple Tumors

Multiple tumors that are individually smaller than the single tumor limit may not have the same negative sequellae as a single tumor. Multiple tumors may be allowed to grow up to 150% of the volume compared with the volume of a single tumor. The size limitation of the diameter of any single tumor (2.0 cm in mice or 4.0 cm in rats) should still be applied. Institutional allowance on permissible sizes of tumors typically will be decreased if the tumors are transplanted into immature or genetically runted mice.

Ascites Produced by Tumors

If tumors are expected to grow with accumulation of fluid in the peritoneal cavity (ascites), rodents must be weighed prior to inoculation and subsequently be weighed at regular intervals. All monitoring should be thoroughly described in the IACUC protocol and based on the expected rate of fluid accumulation. Ascites pressure should be relieved before abdominal distension is great enough to cause discomfort or interfere with normal activity. When the BW exceeds 120% of baseline weight, the animal should be euthanized or the fluid removed ("tapped") from the abdominal cavity. The abdominal tap should be performed by trained personnel using proper aseptic technique, with manual restraint or anesthesia, and by using the smallest needle (18–22 gauge) possible that still permits fluid removal. Certain institutions may have limitations on the number of abdominal taps that can be performed prior to the animal reaching a humane endpoint.

Aspects of this section were adapted from institutional documents (IACUC-CORNELL, 2013, IACUC-UPENN, 2010).

wound management considerations

General anesthesia may be necessary for initial wound assessments in rodents, depending on severity. Wounds typically undergo four stages of healing: inflammatory, debridement, repair, and maturation. Appropriate wound therapy is related to determination of the stage of healing (Langlois, 2004). Many factors have an adverse impact on wound healing of the critical patient, including nutritional status, immune status, concurrent disease (e.g., diabetes mellitus), neoplasia and paraneoplastic syndromes, and location of the wound.

Wound management involves culture and antibiotic sensitivity testing, decontamination (clipping of surrounding hairs after covering wound with a water-soluble lubricant), wound lavage with warm sterile saline, application of antiseptic ointment, and wound debridement.

Topical treatments can be applied, with the realization that the grooming behaviors of rodents may lead to inadvertent ingestion of the substances. However, similar treatments to those listed for murine ulcerative dermatitis (UD) (see relevant section in Chapter 2) can be applied, including Neosporin®, Silvadene®, Preparation H® ointment to stimulate collagen synthesis, aloe vera, CarraVet® wound gel, and sugar and unpasteurized honey (Langlois, 2004, Mathews and Binnington, 2002a, 2002b). Toenails should be trimmed to diminish ability to self-injure by scratching at the wound site.

Ideally, wounds should be closed (whether by suture, steel skin clips/staples, or tissue glue like Vetbond™), taking care to avoid creation of tension across tissues, and protected to avoid disruption of the granulation tissue formation (Hernandez-Divers, 2004). If suture closure is preferred, a suture size of 3–0 thickness or smaller is preferred in rodents; individually packaged commercial suture materials, with attached tapered needles, are commonly used. Rodents will have a propensity to chew at incision sites, often as a manifestation of discomfort; therefore, care should be taken to try to bury the suture line into the subcuticular layer and provide appropriate dosages of analgesics. If skin clips or sutures are placed, they should be removed 10 to 14 days after placement. Restraint collars (see relevant section on this topic) can also be applied during the first 7–10 days after wounds are dressed. Additional comprehensive information about critical care management of wounds is reviewed extensively in other texts (Garzotto, 2009, Langlois, 2004).

references

Aldred, AJ, Cha, MC, and Meckling-Gill, KA. 2002. Determination of a humane endpoint in the L1210 model of murine leukemia. *Contemp Top Lab Anim Sci* 41:24–27.

American College of Laboratory Animal Medicine (ACLAM). 2005. Report of the ACLAM Task Force on Rodent Euthanasia. http://www.aclam.org/Content/files/files/Public/Active/report_rodent_euth.pdf.

American Veterinary Medical Association (AVMA). 2013. AVMA Guidelines for the Euthanasia of Animals: 2013 Edition, pp. 1–102. https://www.avma.org/KB/Policies/Documents/euthanasia.pdf.

Angel, MF, Jorysz, M, Schieren, G, Knight, KR, and O'Brien, BM. 1992. Hair removal by a depilatory does not affect survival in rodent experimental flaps. *Ann Plast Surg* 29:297–298.

Bekkevold, C, Rowland, NE, Robertson, K, Reinhard, M, and Battles, AH. 2012. Physiologic effects of deprivation and restriction of water in CD1 mice: role in research and animal welfare. *J Am Assoc Lab Anim Sci* 51:646.

Bergdall, V, and Green, J. 2004. Equipping the operating room for USDA-covered species. *Lab Anim (NY)* 33:35–38.

Brown, C. 2006a. Restraint collars. Part I: Elizabethan collars and other types of restraint collars. *Lab Anim (NY)* 35:23–25.

Brown, C. 2006b. Restraint collars. Part II: Specific issues with restraint collars. *Lab Anim (NY)* 35:25–27.

Caro, A, Brice, AK, and Veeder, CL. 2011. Investigation into a fire caused by improper surgical preparation and surgical instrumentation use in a protocol-related rodent surgery. *J Am Assoc Lab Anim Sci* 50:745.

Danneman, PJ, Suckow, MA, and Brayton, CF. 2012. *The Laboratory Mouse*, 2nd edition. CRC Press, Boca Raton, FL.

DiBartola, S. 2000. *Fluid Therapy in Small Animal Practice.* Saunders, Philadelphia.

Donnelly, TM, and Walberg, J. 2011. Enlarged mouse preputial glands. *Lab Anim (NY)* 40:69–71.

Dontas, IA, Tsolakis, AI, Khaldi, L, Patra, E, and Lyritis, GP. 2010. Malocclusion in aging Wistar rats. *J Am Assoc Lab Anim Sci* 49:22–26.

Esfandiari, A, and Nowrouzian, I. 2006. Efficacy of polypropylene mesh coated with bioresorbable membrane for abdominal wall defects in mice. *J Am Assoc Lab Anim Sci* 45:48–51.

Finlay, JB, Vonderfecht, S, Griffey, S, Ermel, R, and Adamson, TW. 2012. Examination of the effects of depilatory creams on mice. *J Am Assoc Lab Anim Sci* 51:692.

Flecknell, PA. 2001. Analgesia of small mammals. *Vet Clin North Am Exot Anim Pract* 4:47–56, vi.

Garzotto, CK. 2009. Wound management, Chap. 157. In Silverstein, D, and Hopper, K (eds.), *Small Animal Critical Care Medicine.* Saunders, St. Louis, MO.

Hackett, TB, and Lehman, TL. 2005. Practical considerations in emergency drug therapy. *Vet Clin North Am Small Anim Pract* 35:517–525, viii.

Hawkins, MG, and Graham, JE. 2007. Emergency and critical care of rodents. *Vet Clin North Am Exot Anim Pract* 10:501–531.

Hernandez-Divers, SM. 2004. Principles of wound management of small mammals: hedgehogs, prairie dogs, and sugar gliders. *Vet Clin North Am Exot Anim Pract* 7:1–18, v.

Hoggatt, AF, Hoggatt, J, Honerlaw, M, and Pelus, LM. 2010. A spoonful of sugar helps the medicine go down: a novel technique to improve oral gavage in mice. *J Am Assoc Lab Anim Sci* 49:329–334.

Humane endpoints for animals used in biomedical research and testing. 2000. *Inst Lab Animal Res J* 41(2) [entire issue].

IACUC-CORNELL. 2013. ACU 402.02: Humane Intervention Points. http://www.research.cornell.edu/care/documents/ACUPs/ACUP402.pdf.

IACUC-UPENN. 2008. IACUC Guideline. Carbon Dioxide Euthanasia of Rodents. http://www.upenn.edu/regulatoryaffairs/Documents/CARBON_DIOXIDE_EUTHANASIA_OF_RODENTS_GUIDELINE.pdf.

IACUC-UPENN. 2010. IACUC Guideline. Rodent Tumor Production. http://www.upenn.edu/regulatoryaffairs/Documents/iacuc/guidelines/iacucguideline-rodenttumorproduction.pdf.

IACUC-UPENN. 2011a. IACUC Guideline. Food/Fluid Regulation in Rodents. http://www.upenn.edu/regulatoryaffairs/Documents/iacuc/guidelines/iacucguideline-foodfluidregulationinrodents.pdf.

IACUC-UPENN. 2011b. IACUC Guideline. Humane Intervention and Endpoints for Laboratory Animal Species. http://www.upenn.edu/regulatoryaffairs/Documents/iacuc/guidelines/iacucguideline-humaneendpoints-8%2023%2011.pdf.

IACUC-UPENN. 2013. IACUC Guideline. Experimental Autoimmune Encephalomyelitis and Other Demyelinating Rodent Disease Models. http://www.upenn.edu/regulatoryaffairs/Documents/iacucguideline-eaedemyelinatingrodentmodels.pdf.

Kalishman, J, Elson, K, Acosta, H, and Popilskis, S. 2004. Management of veterinary care and post-operative monitoring of rats in an academic research facility. *Contemp Top Lab Anim Sci* 43:52–53.

Klaphake, E. 2006. Common rodent procedures. *Vet Clin North Am Exot Anim Pract* 9:389–413, vii–viii.

Klaunberg, BA, O'Malley, J, Clark, T, and Davis, JA. 2004. Euthanasia of mouse fetuses and neonates. *Contemp Top Lab Anim Sci* 43:29–34.

Klein, RC. 2008. Fire safety recommendations for administration of isoflurane anesthesia in oxygen. *Lab Anim (NY)* 37:223–224.

Langlois, I. 2004. Wound management in rodents. *Vet Clin North Am Exot Anim Pract* 7:141–167.

Lester, PA, Moore, RM, Shuster, KA, and Myers, DD, Jr. 2012. Anesthesia and analgesia, Chap. 2. In Suckow, MA, Stevens, KA, and Wilson, RP (eds.), *The Laboratory Rabbit, Guinea Pig, Hamster, and Other Rodents*. Academic Press, New York.

Levine, S, and Saltzman, A. 1998. An alternative to overnight withholding of food from rats. *Contemp Top Lab Anim Sci* 37:59–61.

Lichtenberger, M. 2007. Shock and cardiopulmonary-cerebral resuscitation in small mammals and birds. *Vet Clin North Am Exot Anim Pract* 10:275–291.

Lichtenberger, M, and Ko, J. 2007. Critical care monitoring. *Vet Clin North Am Exot Anim Pract* 10:317–344.

Locke, B, Andrutis, KA, and Battles, AH. 2011. Cost-effective solutions to improve surgical technique and asepsis during rodent surgery. *J Am Assoc Lab Anim Sci* 50:734–735.

Madeddu, P, Emanueli, C, Spillmann, F, Meloni, M, Bouby, N, Richer, C, Alhenc-Gelas, F, Van Weel, V, Eefting, D, Quax, PH, Hu, Y, Xu, Q, Hemdahl, AL, van Golde, J, Huijberts, M, de Lussanet, Q, Struijker Boudier, H, Couffinhal, T, Duplaa, C, Chimenti, S, Staszewsky, L, Latini, R, Baumans, V, and Levy, BI. 2006. Murine models of myocardial and limb ischemia: diagnostic end-points and relevance to clinical problems. *Vascul Pharmacol* 45:281–301.

Mader, DR. 2002. Emergency/ICU procedures in small mammals, pp. 608–610. Eighth International Veterinary Emergency and Critical Care Symposium, San Antonio, TX, September 2002.

Mathews, KA, and Binnington, AG. 2002a. Wound management using honey. *Compendium* 24:53–60.

Mathews, KA, and Binnington, AG. 2002b. Wound management using sugar. *Compendium* 24:41–50.

Mercogliano, J, Maher, J, and Lee, J. 2002. Monitoring the effects of repeated blood sampling on ECG, body temperature, and blood pressure using telemetry devices. *Contemp Top Lab Anim Sci* 41:107–108.

Morrisey, JK. 2003a. Practical analgesia and anesthesia of exotic pets, pp. 591–596. Ninth International Veterinary Emergency and Critical Care Symposium, New Orleans, LA, September 2003.

Morrisey, JK. 2003b. Transfusion medicine in exotics, pp. 578–584. Ninth International Veterinary Emergency and Critical Care Symposium, New Orleans, LA, September 2003.

Muir, WW, Hubbell, JAE, and Skarda, R. 1989. Euthanasia, Chap. 27. In Reinhart R (ed.), *Handbook of Veterinary Anesthesia*. Mosby, St. Louis, MO.

Nadon, NL. 2004. Maintaining aged rodents for biogerontology research. *Lab Anim (NY)* 33:36–41.

Narver, HL. 2013. Care and monitoring of a mouse model of melanoma. *Lab Anim (NY)* 42:92–98.

National Research Council. 2011. *Guide for the Care and Use of Laboratory Animals*, 8th edition. National Academies Press, Washington, DC.

Nemzek, JA, Xiao, HY, Minard, AE, Bolgos, GL, and Remick, DG. 2004. Humane endpoints in shock research. *Shock* 21:17–25.

Pacharinsak, C, and Beitz, A. 2008. Animal models of cancer pain. *Comp Med* 58:220–233.

Paster, EV, Villines, KA, and Hickman, DL. 2009. Endpoints for mouse abdominal tumor models: refinement of current criteria. *Comp Med* 59:234–241.

Paul-Murphy, J. 1996. Little critters: emergency medicine for small rodents, pp. 714–718. Fifth International Veterinary Emergency and Critical Care Symposium, San Antonio, TX.

Phillips, PM, Jarema, KA, Kurtz, DM, and MacPhail, RC. 2010. An observational assessment method for aging laboratory rats. *J Am Assoc Lab Anim Sci* 49:792–799.

Pritchett, K, Corrow, D, Stockwell, J, and Smith, A. 2005. Euthanasia of neonatal mice with carbon dioxide. *Comp Med* 55:275–281.

Pritchett-Corning, KR. 2009. Euthanasia of neonatal rats with carbon dioxide. *J Am Assoc Lab Anim Sci* 48:23–27.

Public Health Service. 1996. *Public Health Service Policy on Humane Care and Use of Laboratory Animals*. U.S. Department of Health and Human Services, Washington, DC.

Ray, MA, Johnston, NA, Verhulst, S, Trammell, RA, and Toth, LA. 2010. Identification of markers for imminent death in mice used in longevity and aging research. *J Am Assoc Lab Anim Sci* 49:282–288.

Rowland, NE. 2007. Food or fluid restriction in common laboratory animals: balancing welfare considerations with scientific inquiry. *Comp Med* 57:149–160.

Schenk, M, Orlando, N, Schenk, N, Haung, CA, and Duran-Struuck, R. 2012. Development of a murine hematopoietic tumor model scoring system. *J Am Assoc Lab Anim Sci* 51:631.

Sharp, PE, and Villano, J. 2012. *The Laboratory Rat*, 2nd edition. CRC Press, Boca Raton, FL.

Stanford, M. 2004. Practical use of capnography in exotic animal anesthesia. *Exotic DVM* 6.3:49–52.

Taylor, WM, and Grady, AW. 1998. Catheter-tract infections in rhesus macaques (*Macaca mulatta*) with indwelling intravenous catheters. *Lab Anim Sci* 48:448–454.

Toth, LA, and Gardiner, TW. 2000. Food and water restriction protocols: physiological and behavioral considerations. *Contemp Top Lab Anim Sci* 39:9–17.

Trammell, RA, Cox, L, and Toth, LA. 2012. Markers for heightened monitoring, imminent death, and euthanasia in aged inbred mice. *Comp Med* 62:172–178.

Turner, PV, Albassam, MA, and Walker, RM. 2001. The effects of overnight fasting, feeding, or sucrose supplementation prior to necropsy in rats. *Contemp Top Lab Anim Sci* 40:36–40.

Turner, PV, Pekow, C, Vasbinder, MA, and Brabb, T. 2011. Administration of substances to laboratory animals: equipment considerations, vehicle selection, and solute preparation. *J Am Assoc Lab Anim Sci* 50:614–627.

Workman, P, Aboagye, EO, Balkwill, F, Balmain, A, Bruder, G, Chaplin, DJ, Double, JA, Everitt, J, Farningham, DA, Glennie, MJ, Kelland, LR, Robinson, V, Stratford, IJ, Tozer, GM, Watson, S, Wedge, SR, and Eccles, SA. 2010. Guidelines for the welfare and use of animals in cancer research. *Br J Cancer* 102:1555–1577.

Yan, L, Combs, GF, Jr, DeMars, LC, and Johnson, LK. 2011. Effects of the physical form of the diet on food intake, growth, and body composition changes in mice. *J Am Assoc Lab Anim Sci* 50:488–494.

Zhang, YP, Onifer, SM, Burke, DA, and Shields, CB. 2001. A topical mixture for preventing, abolishing, and treating autophagia and self-mutilation in laboratory rats. *Contemp Top Lab Anim Sci* 40:35–36.

resources and additional information

introduction

Additional resources and helpful references are provided in this chapter for both the general specialty of laboratory animal science and specifics related to clinical laboratory animal medicine.

organizations

Professional organizations that provide clinical laboratory animal medicine information are limited; however, those that exist provide access to resources for their members and promote a network of collaboration between professionals in the field.

American Association for Laboratory Animal Medicine (AALAS)

http://www.aalas.org/

AALAS serves a diverse professional group, ranging from research investigators to animal care technicians to veterinarians. AALAS publishes relevant specialty journals and materials, from which the majority of materials for this text were derived. These publications include the *Journal of*

AALAS (formerly *Contemporary Topics in Laboratory Animal Science*), *Comparative Medicine* (formerly *Laboratory Animal Science*), and *Tech Talk*. AALAS began to publish an additional journal, *Laboratory Animal Science Professional*, in 2013. This organization offers an online comprehensive training module library, the AALAS Learning Library, which offers members a diverse listing of courses relevant to laboratory animal science. The AALAS website provides a link (http://nationalmeeting.aalas.org/past_meeting_abstracts.asp) for members to access those abstracts (dating through 1992) presented at the annual national AALAS meetings; these abstracts provide key suggestions and considerations for laboratory animal care.

American College of Laboratory Animal Medicine (ACLAM)

http://www.aclam.org/

The American College of Laboratory Animal Medicine is comprised of veterinarians certified in the specialty of laboratory animal medicine. This group conducts an annual ACLAM Forum for continuing education to advance the humane care and responsible use of laboratory animals. ACLAM posts published position statements and reports (http://www.aclam.org/education-and-training/position-statements-and-reports) on topics including veterinary care, animal experimentation, pain and distress, rodent surgery, and rodent euthanasia.

American Society of Laboratory Animal Practitioners (ASLAP)

http://www.aslap.org/

ASLAP membership includes veterinary professionals, trainees, and students with an interest in laboratory animal practice. ASLAP supports educational sessions aimed to promote knowledge, ideas, and information for the benefit of animals and society at both the national AALAS and the annual American Veterinary Medical Association (AVMA) conferences.

Institute for Laboratory Animal Research (ILAR)

http://dels.nas.edu/ilar/

The mission of ILAR is to evaluate and disseminate information on issues related to the scientific, technological, and ethical use of animals and related biological resources in research, testing, and education. The organization publishes comprehensive topical issues of the *ILAR Journal* on relevant subjects for the care and use of laboratory animal species and those individuals that work with them. ILAR functions as a component of the National Academies to provide expertise to the federal government, the international biomedical research community, and the public.

publications

Published materials, books, journals, and other documents are extremely valuable resources for clinical laboratory animal information and discussion of relevant experimental models.

Books

Banks, RE, Sharp, JM, Doss, SD, and Vanderford, DA. *Exotic Small Mammal Care and Husbandry*. Wiley-Blackwell, Ames, IA, 2010.

Birchard, SJ, and Sherding, RG. *Saunders Manual of Small Animal Practice*, 3rd edition. Saunders, Philadelphia, 2005.

Danneman, P, Suckow, MA, and Brayton, C. *The Laboratory Mouse*, 2nd edition. CRC Press, Boca Raton, FL, 2012.

Ford, RB, and Mazzaferro, E. *Kirt & Bistner's Handbook of Veterinary Procedures and Emergency Treatment*, 9th edition. Saunders, Philadelphia, 2012.

Fox, JG, Barthold, SW, Davisson, MT, Newcomer, CE, Quimby, FW, Smith, AL. *The Mouse in Biomedical Research*. American College of Laboratory Animal Medicine Series. Elsevier via Academic Press, New York, 2007.

Gaertner, DJ, Hankenson, FC, Hallman, T, and Batchelder, MA. Anesthesia and analgesia in rodents, Chap. 10. In Fish RE, Brown, MJ, Danneman PJ, and Karas, AZ (eds.), *Anesthesia and Analgesia for Laboratory Animals*. Academic Press, San Diego, CA, 2008.

Heinlich, H, Bullock, GR, and Petrusz, P. *The Laboratory Mouse.* Academic Press, San Diego, CA, 2004.

Hrapkiewicz, K, and Medina, L. *Clinical Laboratory Animal Medicine: An Introduction,* 3rd edition. Wiley-Blackwell, Ames, IA, 2006.

Johnson-Delaney, C.A. *Exotic Companion Medicine Handbook for Veterinarians.* Wingers, Lake Worth, FL, 1996.

Macintire, DK, Drobatz, KJ, Haskins, SC, and Saxon, WD. *Manual of Small Animal Emergency and Critical Care Medicine.* Lippincott Williams & Wilkins, New York, 2005.

Murtaugh, RJ, and Kaplan, PM. *Veterinary Emergency and Critical Care Medicine.* Mosby-Year Book, St. Louis, MO, 1992.

National Research Council (NRC). *Recognition and Alleviation of Distress in Laboratory Animals.* National Academies Press, Washington, DC, 2008.

National Research Council (NRC). *Recognition and Alleviation of Pain in Laboratory Animals.* National Academies Press, Washington, DC, 2009.

National Research Council (NRC). *Guide for the Care and Use of Laboratory Animals,* 8th edition. National Academies Press, Washington, DC, 2011.

Oglesbee, BL. *Blackwell's Five-Minute Veterinary Consult: Small Mammal,* 2nd edition. Wiley-Blackwell, Ames, IA, 2011. [*Note:* Contains a section on rodents.]

Percy, DH, and Barthold, SW. *Pathology of Laboratory Animal Rodents,* 3rd edition. Wiley-Blackwell, Ames, IA, 2007.

Plunkett, SJ. *Emergency Procedures for the Small Animal Veterinarian,* 2nd edition. Saunders, Philadelphia, 2000. [*Note:* Contains a section on exotics.]

Pritchett-Corning, KR, Girod, A, Avellaneda, G, Fritz, PE, Chou, S, and Brown, MJ. *Handbook of Clinical Signs in Rodents and Rabbits.* Charles River Laboratories, Wilmington, MA, 2010.

Quesenberry, K, and Carpenter, JW. *Ferrets, Rabbits and Rodents: Clinical Medicine and Surgery,* 2nd edition. Saunders, Philadelphia, 2003.

Sharp, PE, and Villano, J. *The Laboratory Rat,* 2nd edition. CRC Press, Boca Raton, FL, 2012.

Silverstein, D, and Hopper, K. *Small Animal Critical Care Medicine.* Saunders, Philadelphia, 2008.

Suckow, MA, Weisbroth, SH, and Franklin, CL. *The Laboratory Rat*, 2nd edition. ACLAM series. American College of Laboratory Animal Medicine Series. Elsevier via Academic Press, New York, 2006.

Suckow, MA, Stevens, KA, and Wilson, RP. *The Laboratory Rabbit, Guinea Pig, Hamster and Other Rodents*. American College of Laboratory Animal Medicine Series. Elsevier via Academic Press, New York, 2012.

Periodicals

ALN® (Animal Lab News) Magazine [published by Vicon Publishing, Inc.], http://www.alnmag.com/

Comparative Medicine [published by AALAS] (formerly *Laboratory Animal Science*)

ILAR Journal [published by Oxford Journals], http://ilarjournal.oxfordjournals.org/

Journal of the American Association for Laboratory Animal Science [published by AALAS] (formerly *Contemporary Topics in Laboratory Animal Science*)

Journal of Exotic Pet Medicine [published by Elsevier, Inc.], http://www.exoticpetmedicine.com/home

Lab Animal magazine [published by Nature Publishing Group], http://www.labanimal.com/laban/index.html

Laboratory Animals [published by the Royal Society of Medicine journals], http://la.rsmjournals.com/

Laboratory Animal Science Professional [published by AALAS]

Tech Talk [published by AALAS]

Veterinary Clinics of North America—Exotic Animal Practice [published by Elsevier, Inc.], http://www.vetexotic.theclinics.com/

electronic resources

AALAS Learning Library

https://www.aalaslearninglibrary.org/default.asp

The AALAS Learning Library provides training modules of benefit for technicians, veterinarians, managers, IACUC members, and investigators working with animals in a research or educational setting.

Animal Care Training Services (ACTS)

http://actstraining.com/

ACTS is a training company that specializes in the daily operations of lab animal research institutions. They provide "job-specific skill training" program modules, assist with training of staff to achieve AALAS certification levels for technicians, and support on-site seminars on a variety of customized training topics at local and regional sites.

AVMA Guidelines for the Euthanasia of Animals: 2013 Edition

https://www.avma.org/KB/Policies/Documents/euthanasia.pdf

The 2013 guidelines, established by membership of the Panel on Euthanasia, set criteria for euthanasia, specified appropriate euthanasia methods and agents, and are intended to assist veterinarians. In this version, methods, techniques, and agents of euthanasia have been updated, and detailed descriptions have been included to assist veterinarians in applying their professional and clinical judgment.

CompMed™ listserv

http://www.aalas.org/online_resources/listserves.aspx#compmed

CompMed is an e-mail list for discussion of comparative medicine, laboratory animals, and topics related to biomedical research. CompMed is limited to participants who are involved in some aspect of biomedical research or veterinary medicine, including veterinarians, technicians, animal facility managers, researchers, and graduate/veterinary students. AALAS membership is not required to subscribe to this group.

To subscribe:

Send e-mail to: LISTSERV@LISTSERV.AALAS.ORG

Message body: SUBSCRIBE COMPMED Yourfirstname Yourlast name (Example: SUBSCRIBE COMPMED John Doe)

Drug Enforcement Agency (DEA), Office of Diversion Control

http://www.deadiversion.usdoj.gov/index.html

This website provides information about registering to obtain licensure for appropriating controlled substances (drugs) for use in veterinary medicine in the United States and its territories.

IACUC-Forum listserv

http://www.aalas.org/online_resources/listserves.aspx#IACUC-Forum

IACUC-Forum is a member benefit for current AALAS institutional members. There are no fees for this service; it is included as part of institutional membership dues. Current institutional contact persons may enroll their IACUC members and IACUC staff on IACUC-Forum; the IACUC members and IACUC staff who have access to the list are not required to be members of AALAS for the purposes of this list. Only individuals directly related to the IACUC are eligible to have access to the list.

To subscribe, complete and submit the application form found on the web link.

International Council for Laboratory Animal Science (ICLAS)

http://iclas.org/

ICLAS is the international scientific organization dedicated to advancing human and animal health by promoting the ethical care and use of laboratory animals in research worldwide. From the ICLAS membership page (http://iclas.org/members/member-list), the following international laboratory animal science groups can be accessed:

Asociación Argentina de Ciencia y Tecnología de Animales de Laboratorio (AACyTAL)

Asociación Chilena de Ciencias del Animal de Laboratorio (ASOCHICAL)

Asociacion Mexicana de la Ciencia de los Animales de Laboratorio (AMCAL)

Asociación Uruguaya de Ciencia y Tecnología de Animales de Laboratorio (AUCyTAL)

Association Française des Sciences et Techniques de l' Animal de Laboratoire (AFSTAL)

Associations of Central America, Caribbean and Mexico Laboratory Animal Science (ACCMAL)

Associazione Italiana per le Scienze degli Animali da Laboratorio (AISAL)

Australia and New Zealand Laboratory Animal Association (ANZLAA)

Belgian Council for Laboratory Animal Science (BCLAS)

Canadian Association for Laboratory Animal Science (CALAS/ ACSAL)

Chinese Association for Laboratory Animal Science (CALAS, China)

Chinese–Taipei Society of Laboratory Animal Sciences (CSLAS)

Finland Laboratory Animal Science (FinLAS)

German Society for Laboratory Animal Science (GV-SOLAS)

Israeli Laboratory Animal Forum (ILAF)

Japanese Association for Laboratory Animal Science (JALAS)

Japanese Society for Laboratory Animal Resources (JSLAR)

Korea Research Institute of Bioscience and Biotechnology (KRIBB)

Korean Association for Laboratory Animal Science (KALAS, Korea)

Laboratory Animal Science Association (LASA, United Kingdom)

Laboratory Animal Scientist's Association (LASA, India)

Laboratuvar Hayvanları Bilimi Derneği (TURKEY)/ LASA-Turkey

Nederlandse Vereniging voor Proefdierkunde, (NVP)/ Biotechnische Vereniging (BV)

Scandinavian Society for Laboratory Animal Science (SCAND-LAS)

Sociedad Española Para las Ciencias del Animal De Laboratorio, (SECAL)

Sociedade Brasileira de Ciencia de Animais de Laboratorio (SBCAL)

South-African Association for Laboratory Animal Science (SAALAS, South Africa)

Swiss Laboratory Animal Science Association–Schweizerische Gesellschaft für Versuchstierkunde (SGV)

Thai Association for Laboratory Animal Science (TALAS, Thailand)

International Mouse Strain Resource (IMSR)

http://www.findmice.org/

The IMSR is a multi-institutional international collaboration supporting the use of the mouse as a model system for studying human biology and disease. The primary goal of the IMSR is to provide a web-searchable catalog that will assist the international research community in finding the mouse resources needed.

The IMSR began with an initial collaboration between the Mouse Genome Informatics (MGI) group at the Jackson Laboratory and the Medical Research Council Mammalian Genetics Unit at Harwell, United Kingdom. Many institutions and collaborators are now contributing mouse resource information to the IMSR catalog.

Mouse Genome Database

http://www.informatics.jax.org/

The U.S. National Institutes of Health provide support for this reference database maintained through the website of the Jackson Laboratory. This database provides a resource for mouse genetic, genomic, and biological information, such as gene characterization, characteristics of inbred strains, descriptions of mutant phenotypes, and additional related subjects.

National Institutes of Health Office of Research Infrastructure Programs (ORIP): Rodent Resources

http://dpcpsi.nih.gov/orip/cm/rodents_index.aspx

ORIP's laboratory rodents program funds development of genetically engineered rodents and research rodent colonies, facilities that distribute rodents and related biological materials, and new ways to study, diagnose, and eliminate laboratory rodent disease. Related links from this page include the following:

Rodent Resources for Researchers, a listing of hyperlinks to various mutant mouse resource centers, phenotyping programs, mutant rat resources, and resources for rat genetic models (http://dpcpsi.nih.gov/orip/cm/rodent_resource_researchers.aspx).

Office of Laboratory Animal Welfare

http://grants.nih.gov/grants/olaw/olaw.htm

The Office of Laboratory Animal Welfare (OLAW) provides guidance and interpretation of the Public Health Service (PHS) Policy on Humane Care and Use of Laboratory Animals, supports educational programs, and monitors compliance with the policy by assured institutions and PHS funding components to ensure the humane care and use of animals in PHS-supported research, testing, and training, thereby contributing to the quality of PHS-supported activities. The site contains an extensive listing of answers to frequently asked questions, providing further commentary on topics related to research animal welfare (e.g., pharmaceutical-grade drug definitions, euthanasia, housing expectations per the National Resource Council's 2011 *Guide for the Care and Use of Laboratory Animals* [National Academies Press, Washington, DC]).

Pubmed

http://www.ncbi.nlm.nih.gov/pubmed

PubMed is an electronic database supported by the U.S. National Library of Medicine and National Institutes of Health; it comprises more than 22 million citations for biomedical literature from MEDLINE, life science journals, and online books.

Citations may include links to full-text content from PubMed Central and publisher websites.

Rat Genome Database

http://rgd.mcw.edu/

The Rat Genome Database is a collaborative effort between leading research institutions involved in rat genetic and genomic research. This resource is monitored and supported by grant HL64541. "Rat Genome Database," awarded to Dr. Howard J. Jacob at the Medical College of Wisconsin by the National Heart Lung and Blood Institute (NHLBI) of the National Institutes of Health (NIH). The Rat Genome Database was created to serve as a repository of rat genetic and genomic data, as well as mapping, strain, and physiological information. It also facilitates investigators' research efforts by providing tools to search, mine, and analyze these data.

TechLink listserv

http://www.aalas.org/online_resources/listserves.aspx

TechLink is an electronic mailing list (listserve) created especially for animal care technicians in the field of laboratory animal science. Open to any AALAS national member, TechLink serves as a method for laboratory animal technicians to exchange information and conduct discussions of common interest via e-mail messages with technicians in the United States and other countries around the world.

To subscribe:

Send e-mail to: LISTSERV@LISTSERV.AALAS.ORG

Message body: SUBSCRIBE TECHLINK Yourfirstname Yourlastname

(Example: SUBSCRIBE TECHLINK John Doe)

Veterinary Bioscience Institute (VBI)

http://www.vetbiotech.com/

VBI offers training modules for experimental and veterinary surgical and biomethodology training for technical and medical staff. VBI provides online training with hands-on training modules for a variety of rodent surgical procedures, including innovative approaches like laparoscopy. The site provides fee-for-service access to webinars and learning modules online.

Veterinary Emergency and Critical Care Society (VECCS)

http://www.veccs.org/

VECCS aims to raise the level of patient care for seriously ill or injured animals through quality education and communication programs. The society works closely with the American College of Veterinary Emergency and Critical Care (ACVECC) to provide information related to life-threatening and acute disease conditions in pet medicine.

Veterinary Information Network (VIN)

http://www.vin.com/VIN.plx

VIN serves as an online resource for veterinarians with content submitted by veterinarians from various specialties in clinical practice. Membership to the site, which supports conference proceedings from a variety of veterinary annual conferences, is for a fee; however, veterinary students and academicians are allowed access at no charge.

Topics of interest can be searched for input from colleagues, and continuing education courses and lectures are available.

commercial resources

ALN Buyer's Guide

http://www.alnmag.com/buyers-guide [also available in hard-copy format]

The ALN® Magazine and ALN World™ Buyer's Guides are comprehensive sources of resources, products, and information to design, build, and equip animal research facilities. Direct links are provided to vendor and commercial information concerning products for laboratory animals; animal care and maintenance; facility design, materials, and equipment; laboratory and research equipment and supplies; organizations; surgical and medical equipment and supplies; veterinary and research services; and other services, materials, and equipment.

Lab Animal Buyer's Guide

http://guide.labanimal.com/guide/index.html [also available in hard-copy format]

A comprehensive database of suppliers, products, and services in laboratory animal care. The source can be searched by the name of a specific supplier or product category. Searchable categories include animal care, animals, food and water, housing, husbandry, information resources and management, plant, research, services, surgery, and veterinary medical care.

appendix A: glossary of acronyms and terms

Abbreviation/Word	Definition
Ad libitum	Continuous access to food and fluid sources
Autochthonous	Tumor burden originating within the host animal
Ascites	Fluid in the peritoneal cavity
BAR	Bright, alert, responsive
BCS	Body condition score
BID	Twice daily treatment
BT	Body temperature
BUN	Blood urea nitrogen
BW	Body weight
CFA	Complete Freund's adjuvant
CO_2	Carbon dioxide
CS	Controlled substance
DEA	Drug Enforcement Agency; provides registrations to entities that intend to use any CS in the United States and its territories
DMSO	Dimethyl sulfoxide; solvent with anti-inflammatory properties, typically used as a drug carrier
EAE	Experimental allergic encephalitis
ECG (EKG)	Electrocardiogram
Ectopic	Site of tumor growth different from the tissue of origin (i.e., liver tumor cells transplanted under the renal capsule); opposite of orthotopic
EDTA	Ethylenediaminetetraacetic acid
EMLA®	Eutectic Mixture of Local Anesthetics (topical eutectic mixture of 2.5% prilocaine and 2.5% lidocaine cream)
EOD	Every other day
Erythema	Reddened skin, may be thickened
ET	Endotracheal

(*Continued*)

Abbreviation/Word	Definition
Fasting	Food access is removed, yet animals have ad libitum access to fluid (i.e., water)
FNA	Fine-needle aspirate
g	Gram (unit of weight)
GY	Gray (unit of radiation)
HCT	Hematocrit
IACUC	Institutional Animal Care and Use Committee
ICU	Intensive care unit
ID	Intradermal
IM	Intramuscular
IP	Intraperitoneal
IT	Intratracheal
IV	Intravenous
LRS	Lactated Ringer's solution
Metastasis	Spread of tumor cells from primary site to distant sites in the body
MUS	Mouse urologic syndrome
NOD	Nonobese diabetic (model for type 1 diabetes)
NSAID	Nonsteroidal anti-inflammatory drug
Orthotopic	Anatomically correct site for tumor transplantation (i.e., liver tumor cells transplanted into the liver); opposite of ectopic
PE	Polyethylene
PO	Per os (by mouth)
Restriction (of food/fluid)	Total volume of food or fluid is strictly monitored and controlled
RO	Retro-orbital
SC	Subcutaneous
Scheduling (of food/fluid)	Animal consumes as much food or fluid as desired at regular intervals
SCID	Severe combined immunodeficiency (mutation)
SID	Once daily treatment
Syngeneic	Tumor cells transplanted between animals of same inbred strain
TBI	Total body irradiation
TBV	Total blood volume
UD	Ulcerative dermatitis
Ulceration	Circumscribed, inflamed, and "open" skin lesion with death (necrosis) of surrounding tissues
Xenogeneic	Tumor cells transplanted between different species of animals (i.e., human cells transplanted into a mouse)

appendix B: suggested medical supplies for rodent critical care

- Alcohol swabs
- Antiseptics (Betadine [povidone-iodine] swabs, chlorhexidine solution)
- Bacterial culturettes/blood culture medium
- Blood analyzer (portable hand-held or table-top)
- Blood collection tubes (red top, green top [heparin], purple top [EDTA])
- Catheters (IV)
- Cotton-tipped applicators (single use, sterile) for topical ointment applications
- Disposable hypodermic needles (23 to 26 gauge for size range)
- Disposable syringes (1 to 3 ml)
- Endotracheal tubes (uncuffed 1.0–2.0 mm for rodents > 100 g)
- Epsom salts (to treat pododermatitis, etc.)
- Feeding needles (for orogastric gavage; 22 gauge, ball tipped)
- Fluids (see Chapter 4)
- Fluorescein stain
- Gauze (4 × 4) sponges
- Glucometer
- Lanolin ointment

- Meloxicam
- Nail clippers for teeth and nail trimming (rats)
- Nose cones for anesthesia
- Nutritional supplements (see Chapter 4)
- Ophthalmoscope
- Otoscope
- Refractometer
- Scalpel blades/handles
- Scissors for teeth and nail trimming (mice)
- Silver sulfadiazine ointment
- Stethoscope (pediatric)
- Surgery instrument packs
- Surgical draping materials (see Chapter 4)
- Suture with attached needles
- Tape
- Tissue glue
- Topical antibiotic ointment ± steroid
- Tweezers
- Vitamin E ointment
- Warm-water recirculating blankets

appendix C: rodent formulary

introduction

Selection of appropriate drug and therapeutic regimens requires careful consideration of multiple factors, including published adverse effects, to maximize effectiveness and minimize risks. Consideration of the selected species, the intended procedure, and the practicality of available agents contributes to the choice of treatments utilized in a given clinical case. Procedures in animals that may cause more than slight pain or distress should be performed with appropriate sedation, analgesia, or anesthesia; one should assume that any procedures deemed painful to humans are therefore able to cause pain to animals (Interagency Research Animal Committee [IRAC], 1985). Pharmaceutical-grade drugs should be used, whenever available, for animal procedures. These agents are defined in detail by the Office of Laboratory Animal Welfare (see relevant section in Chapter 5); as well, expired agents may not be used in any laboratory animals (National Research Council [NRC], 2011).

Dosing in rodents is typically off label and will vary depending on age, gender, strain, and condition of the animals. Pregnant rodents will require special consideration depending on the stage of pregnancy, whether the agent under consideration crosses the placenta, and whether potential effects on the fetus will alter experimental data. Avoiding drug problems during therapy of the critical patient takes preplanning and foresight (Hackett and Lehman, 2005, Meador, 1998). Guidance is available for review of drug interactions, adverse effects, and indications and contraindications for multimodal therapy or "balanced anesthesia" in veterinary patients (Flecknell, 2001, Plumb, 2005). Balanced anesthesia is defined as the "administration of a mixture of sedatives, analgesics, and anesthetics to produce anesthesia with lower doses than would be necessary if each component were used individually" (He et al., 2010 p. 45). It will be critical to obtain an accurate body weight (BW) for each animal prior to administration of a calculated drug dose to limit the potential of adverse effects related to either over-or underdosing.

The following abridged formulary is adapted from numerous rodent references (Danneman et al., 2012, Gaertner et al., 2008, IACUC-UPENN, 2010, Oglesbee, 2011).

induction agents

Induction agents and premedications can calm the patient, smooth anesthetic induction and recovery, and reduce the dose of anesthetic agent needed. Preemptive analgesia should be administered with the induction agents.

Induction Agent	Species	Dosage (mg/kg) (Unless Specified)	Route of Administration
Atropine	Rodents	0.05	SC, IP, IV
Diazepam	Rodents	1–3	IP, SC
Isoflurane	Mice	4%	Inhaled
	Neonatal mice	2–4%	Inhaled
	Rats	5%	Inhaled
Midazolam	Rodents	0.5–2.0	IP, SC
Propofol	Mice	26	IV
	Rats	10	IV

IP = intraperitoneal; IV = intravenous; SC = subcutaneous.

anesthetics

Anesthesia should be provided to animals undergoing procedures that cause more than momentary or slight pain or distress. Anesthetics render the animal unconscious without loss of vital functions. *Inhalant anesthetics* provide a reliable and reversible means of rendering rodents unconscious in order to perform surgeries and other intricate or potentially painful procedures. Injectable anesthetics may not be as predictable in efficacy between animals;

however, they are documented to provide sedation and even anesthesia at a surgical plane. For those drugs that are controlled substances (designated in the formulary tables by CS), and if used in the United States and its territories, a Drug Enforcement Administration (DEA) license will be required to obtain these drugs for use in animals.

Anesthetic	Species	Dosage (mg/kg) (Unless Specified)	Route of Administration
Chloral hydrate	Mice	370–400	IP
	Rats	300–450	IP
	Rats	400–600	SC
Hypothermia (neonatal pups only)	Rodent pups	Placed on crushed ice, separated by a thin layer to avoid direct contact with ice	Contact
Isoflurane	Mice	0.08–1.5%	Inhaled
	Neonatal mice	0.25–2.5%	Inhaled
	Rats	0.25–2.5%	Inhaled
Ketamine (CS)/ diazepam (CS)	Mice	100 K/5 D	IP
	Rats	40 K/5 D	IP
Ketamine (CS)/ midazolam (CS)	Mice	50–75 K/1–10 M	IP
	Rats	60 K/0.4 M	IP
Ketamine (CS)/ xylazine	Mice	90–150 K/7.5–16 X	IP
	Rats	40–80 K/5–10 X	IM, IP
Ketamine (CS)/ xylazine/ acepromazine (necessary for surgical plane of anesthesia)	Mice	70–100 K/5–10 X/1–3 A	IP
	Rats	40 K/8.0 X/4.0 A	IM, IP
Medetomidine/ fentanyl (CS)	Rats	200–300 μg/kg M/300 μg/kg F	IP
Sevoflurane	Rats	2–2.4%	Inhaled
Sodium pentobarbital (CS)	Mice	30–90	IP
	Rats	30–60	IP
Tiletamine (CS)/ zolazepam (CS)	Rats	20–40	IP
Thiobarbital (Inactin) (CS)	Mice	80	IP
Tribromoethanol (TBE or Avertin)	Mice	125–300	IP
Tribromoethanol/ medetomidine	Rats	150 T/0.5 M (reversal 2.5 mg/kg atipamezole)	IP

CS = controlled substance; IM = intramuscular; IP = intraperitoneal; IV = intravenous; SC = subcutaneous.

analgesics

Analgesia should be provided to animals undergoing procedures that cause more than momentary or slight pain or distress. Analgesics reduce or relieve pain without loss of consciousness. Systemic or local analgesics may also reduce the anesthetic requirements and have a preemptive effect on pain perception that persists into the recovery period. Preemptive, but also immediate postoperative, analgesic administration is important for adequate pain relief in postsurgical rodents.

Note: *Please see further published considerations about specific analgesics at the end of the formulary tables.*

Analgesics	Species	Dosage (mg/kg) (Unless Specified)	Route of Administration
Acetaminophen (Tylenol®)	Rodents	100–300	PO, SC, IP
Aspirin (acetyl salicylic acid) (administer every 4–24 h)	Rats	50–100	PO
	Mice	50–100	PO
	Rodents	20	SC
	Rodents	100–120	IP
Buprenorphine (Buprenex®) (CS) (administer every 6–8 h)	Mice	0.5–2.0	SC,IP
	Rats	0.01–0.10 May need doses up to 5.0–10.0 mg/kg if dosed orally	SC, IP, IM
Butorphanol (CS)	Mice	1–5	SC
	Rats	1–5	SC
Carprofen (Rimadyl®) (administer at least every 6–12 h)	Mice (for acute incisional pain)	5	SC
	Rats	5–15	PO, SC
Celecoxib	Rats	10–20	PO
Clonidine	Mice	0.25–0.5	PO
	Mice	0.001–0.1	IP
Diclofenac	Mice	9.0–28	IP
Dipyrone	Rats	50–600	SC, IP, IV
Dipyrone/ morphine (CS)	Rats	177–600 D/3.1–3.2 M	SC, IV
Fentanyl (CS)	Mice	0.025–0.6	SC
	Rats	0.01–1.0	SC
	Rats	2.0–4.0 g/day	PO

(Continued)

Analgesics	Species	Dosage (mg/kg) (Unless Specified)	Route of Administration
Flunixin meglumine (Banamine®)	Mice	4.0–11	SC
	Rats	2.5 every 12–14 h	SC
Ibuprofen (Advil®), Motrin®, Nuprin®)	Mice	40–100	PO, SC
Ibuprofen/ hydrocodone (CS)	Rats	200 I/2.3 H	PO, SC
Ketoprofen (Ketofen®)	Rats	5–15 every 12–24 h	SC, IP
	Mice	5 every 24 h	SC, IP
Lidocaine (Xylocaine®)	Rats	0.67–1.3 mg/ kg/h CRI	SC-pump
Lidocaine/ buprenorphine (CS)	Mice	0.44 m*M* L/0.18 m*M* B in DMSO	Topical
Meloxicam (Metacam®) (administer once daily)	Mice	5.0	SC
	Mice	5.0 (oral suspension)	PO
	Rats	2.0	SC
Morphine (CS)	Mice	10	SC
	Mice	6.1 m*M* in DMSO	Topical
	Rats	2.0–10	SC
	Rats	2.8	SC-Liposome
Naproxen/ hydrocodone (CS)	Rats	200 N/1.3 H	SC
Oxymorphone (CS)	Mice	0.2–0.5	SC-Liposome
	Rats	0.1	IV
		1.2–1.6	SC-Liposome
Tramadol (administer every 12 h)	Rats, mice	5–12.5	SC, IP

CRI = constant rate infusion; CS = controlled substance; DMSO = dimethyl sulfoxide; IP = intraperitoneal; IV = intravenous; PO = by mouth; SC = subcutaneous.

local and topical anesthetics

Local anesthetics can reduce the perception of pain at the surgical site; these should be infiltrated around the incision site prior to recovery of the animal from general anesthesia and preferably prior to initiation of the incision. In conjunction with other agents, their use may allow reduced levels of general anesthetics, which may speed recovery and minimize potential for adverse outcomes.

Local or Topical Anesthetics	Species	Dosage (mg/kg) (Unless Specified)	Route of Administration
Bupivacaine	Rodents	Local infiltration up to 5 mg/kg	Local SC around incision site
EMLA® (may take 30 min for effect)	Rodents	Application of a layer up to 1-mm thick	Topical application
Lidocaine	Rodents	Local infiltration up to 10 mg/kg	Local SC around incision site
Proparacaine (0.5%)	Rodents	1–2 drops per eye	Topical onto eye prior to staining the cornea or prior to RO sampling

Note: Lidocaine and bupivacaine may be mixed at a 1:1 ratio for subcutaneous instillation at the incision site.

RO = retro-orbital; SC = subcutaneous.

reversal agents

Reversal of certain drugs leads to early termination of anesthesia, which may reduce adverse events and allow rapid return of the rodents to the home cage environment. If reversal agents are used, both the anesthetic and the analgesic properties of the drug may be terminated; thus, alternative sources of analgesia should be provided.

Reversal Agents	Species	Dosage (mg/kg) (Unless Specified)	Route of Administration
Yohimbine	Mice	0.5–1.0	SC, IP
Atipamezole (Antisedan®)	Rats	0.5	SC
Naloxone	Rodents	20	IP

IP = intraperitoneal; SC = subcutaneous.

antibiotics

Antibiotics should be selected based on sensitivity and culture results, when available. Typically, more common and broad-spectrum drugs are preferred to begin the treatments prior to culture results.

Antibiotics	Species	Dosage (mg/kg) (Unless Specified)	Route of Administration
Amoxicillin/clavulanic acid (Clavamox®)	Mice	12.5–15 every 12 h	PO
	Rats	150 every 12 h	IM

(Continued)

Antibiotics	Species	Dosage (mg/kg) (Unless Specified)	Route of Administration
Azithromycin	Rodents	10–30 every 24 h	PO
Chloramphenicol	Mice	30–50 every 8h	PO
Enrofloxacin	Rodents	5–10 every 12h Dilution in 300-ml water bottle with 1 mg/ml drug = 300 mg/300-ml bottle	PO, SC
Gentamicin	Rodents	5–10 every 8–12 h	SC, IM
Penicillin	Rodents	22,000–100,000 IU/kg daily	SC
Polymyxin B sulfate-neomycin sulfate-bacitracin zinc (Neosporin® or similar antibiotic ointment)	Rodents	Over-the-counter formulations; apply every 12–24 h to cover the affected area	Topical
Trimethoprim-sulfa	Rodents	15–30 every 12 h Dilute in water source	PO in drinking water

IM = intramuscular; PO = by mouth; SC = subcutaneous.

further commentary on specific anesthetics and analgesics

Multimodal or balanced anesthesia and analgesia (Parker et al., 2011) are the veterinary standards for procedures that are invasive, penetrate a major body cavity, or are predicted to result in moderate-to-severe intensity of impact on the animal. The ideal analgesic regimen will manage patient pain without creation of unwanted side effects or bias to the research model outcomes.

Specific commentary about published concerns and considerations for various analgesic regimens in laboratory rodents are listed alphabetically, with species designations as indicated.

Acetaminophen

Rats

- Acetaminophen should not be dosed in rats above 300 mg/kg PO due to potential for hepatic necrosis and impact on research studies (Hausamann et al., 2002).

- Rats should be acclimated to the novel taste prior to actual administration of flavored suspensions to avoid dramatic reductions in fluid intake due to neophobic tendencies (Bauer et al., 2003, Speth et al., 2001).

Bupivacaine

- Bupivacaine may sting on injection and infusion around the planned incision site; therefore, it should be injected *after* the patient is anesthetized. It should provide pain management at the site of injection for up to 4–6 h.

Buprenorphine

Mice

- Buprenorphine is appropriate for management of acute incisional pain at doses of 0.5 to 2.0 mg/kg SC (Yamada et al., 2009).
- This drug can influence behavior (when dosed up to 1.0 mg/kg) and lead to increased spontaneous locomotor activity, which may adversely affect research outcomes (Cowan et al., 1977).
- In comparative experiments, when mice were dosed (2.0 mg/kg SC) before ova implantation surgery (whereby incisions were made over the flank area with ovary isolation and retraction) and then dosed twice at 6-h intervals after surgery, buprenorphine did not offer superior pain relief compared to one dose of drug given presurgically. Postoperative heart rate and blood pressure parameters were recorded telemetrically and found to have no significant differences between the three doses versus one dose. However, those animals that were given three doses had significant weight loss due to diminished food consumption, which was deemed to be an adverse outcome of the study (Goecke et al., 2005).
- Following **intraperitoneal surgery under isoflurane**, mice have been shown to better tolerate recovery with the addition of a line block at the incision site (bupivacaine, lidocaine, etc.). Buprenorphine can be given intraoperatively at 1.0–2.0 mg/kg IP, then administered twice daily for day 1 after abdominal surgery and subsequently replaced with meloxicam at 5 mg/kg given SID on days 2–3 postprocedure.

- Analgesic efficacy of sustained-release buprenorphine (Bup-SR) dosed at 1.0 mg/kg has been shown to last at least 12 h (compared to 3–5 h for injectable buprenorphine HCl) in male BALB/cJ and SWR/J mice (Carbone et al., 2012).

Rats

- Buprenorphine can increase activity when dosed at 0.1–3.0 mg/kg and lead to abnormal behaviors like repetitive licking and biting of limbs and biting of aspects of the caging environment, along with incidents of fighting, at 4 to 5 h postadministration (Cowan et al., 1977).
- Respiratory depression has been noted in conscious rats following injections (Cowan et al., 1977).
- Oral dosing of buprenorphine is discouraged in rats and has been shown only to be effective for 6- to 8-h intervals for mild-to-moderate pain levels (assessed by hot-water tail-flick assays) at doses approaching 5 mg/kg PO (Gades et al., 2000, Martin et al., 2001, Thompson et al., 2004).
- Buprenorphine offered in flavored gelatin is not readily consumed by rats at doses (5.0–10.0 mg/kg) necessary to induce significant increases in pain threshold, which necessitates orogastric administration by gavage (Martin et al., 2001). For rats undergoing flank laparotomy, 0.3 mg/kg in gelatin provided analgesia and limited postprocedural anorexia and weight loss (Flecknell, Roughan, et al., 1999).
- In a multimodal regimen, specifically for hypophysectomy surgery in rats, following anesthesia with pentobarbital (30–50 mg/kg), animals were provided with buprenorphine (0.05 mg/kg), carprofen (5 mg/kg SC), and fluid therapy (30 ml/kg).
- Rats injected subcutaneously with a 1.2-mg/kg sustained-release formulation (Bup-SR) were tested in thermal nociception and surgical postoperative pain models. In both, Bup-SR showed evidence of analgesia for 2 to 3 days (Foley et al., 2011).
- For management of visceral pain in rats, buprenorphine is less effective than oxymorphone (Gillingham et al., 2001).
- Buprenorphine administration in rats has been linked to side effects of weight loss (Brennan et al., 2009) followed by hyperphagia and weight gain due to pica (Clark et al., 1997, Thompson et al., 2004).

- Pica, manifesting typically as ingestion of large amounts of bedding materials within 1–2 days following buprenorphine administration, has been noted especially in Sprague Dawley rats. Pica can lead to gastric and esophageal impaction, may indicate nausea in rats, and is potentially analogous to emesis in other species (Bender, 1998, Bosgraaf et al., 2004, Clark et al., 1997, Takeda et al., 1993). Pica can also lead to esophageal obstruction, or choking, and may require lavage with hydropropulsion using endoscopy via the mouth to best attempt to advance the obstruction into the stomach (Ovadia and Zeiss, 2002).

Carprofen

Carprofen should be considered as an adjunctive therapy to refine analgesic regimens for rodent surgery and to improve postoperative care (diminish instances of ataxia, bleeding, and weight loss); to increase survival rates; and to maintain animal welfare (Weiner et al., 2011).

Mice

- Carprofen is appropriate for management of acute incisional pain at doses of 5 mg/kg SC every 6 h (Yamada et al., 2009).
- Postlaparotomy, mice can be administered carprofen (5 mg/kg subcutaneously, twice daily for 3 days) with prophylactic antibiotics, like enrofloxacin (30 mg/kg SC SID for 4 days). Some reports have noted that 5 mg/kg is the minimum dose, and that doses up to 10–20 mg/kg carprofen may provide a more effective analgesic dose for mice (Clark et al., 2002).

Rats

- In a multimodal regimen, specifically for hypophysectomy surgery in rats, following anesthesia with pentobarbital (30–50 mg/kg), animals were provided with buprenorphine (0.05 mg/kg), carprofen (5 mg/kg SC), and fluid therapy (30 ml/kg).
- For rats undergoing laparotomy, carprofen (5 mg/kg SC) minimized a postoperative reduction in food and water consumption; however, if dosed orally, higher dose rates should be provided (Flecknell, Roughan, et al., 1999).

Fentanyl

- Transdermal delivery of fentanyl for analgesia has benefits, including more consistent systemic concentrations, reduced dosing frequency, and reduced handling stress. The choice of application site is influenced by the ability of the animal to remove the patch, difficulty of maintaining skin contact by the presence of hair or movement of the animal, and interference with the medical or surgical procedure being performed.
- The interscapular region is a common application site; however, drawbacks of this location include the need to shave the area (which may impair skin integrity) and movement of skin in the conscious animal.

Hypothermia

Mice

- Neonatal rodents typically are resistant to inhalant anesthesia and may best be anesthetized using hypothermia, in essence by placement of altricial pups on ice, separated from direct contact by a thin layer of plastic wrap, parafilm, or paper towel (Phifer and Terry, 1986). Neonates can remain exposed to the ice for 3–10 min to induce torpor for injections or sampling.
- Following the procedure, pups should be slowly rewarmed using a heating source (e.g., incubator ~33°C) or through manual warmth and gentle stimulation and then returned to maternal dams for care. With rewarming, pups become active and responsive within 20–30 min.

Rats

- Anesthesia of neonatal rats (12–14 days old) using hypothermia (by placement of pups on a draped ice pack) combined with inhalant isoflurane anesthesia has been shown to be more effective for subcutaneous implantation procedures than anesthesia with injectable agents, like ketamine and xylazine (dosed 100 and 10 mg/kg, respectively). Rat pups have been documented to crawl back to the dam and commence suckling after hypothermic anesthesia (Libbin and Person, 1979); they are readily accepted by the dam with no long-term side effects noted (Molloy et al., 2004).

Ibuprofen

Mice

- Mice prefer the palatability of oral ibuprofen liquid-gel (at 1 mg/ml) over children's berry-flavored ibuprofen elixir (at 1 mg/ml) as determined in a study in which mice with various size wounds were given either of the two nonsteroidal anti-inflammatory drug (NSAID) options and were further monitored over a 9-day period. Mice consumed significantly more of the liquid-gel formulation. In addition, the mice on liquid-gel consumed twice the amount of food and were more alert, active, and groomed than when given the elixir formulation (Ezell et al., 2012).
- Commercially available cherry-flavored ibuprofen elixir (at 2 mg/ml concentration) has been shown to promote postsurgical recuperation in mice; however, mice may consume this fluid solution in excess of normal and to the detriment of food intake (Bosgraaf et al., 2006).

Ketamine Plus Xylazine

Mice

- Body temperature in rodents may decrease by several degrees following administration of ketamine plus xylazine, and this decrease may be exacerbated by increased urination, defecation, and salivation (Wixson and Smiler, 1997, Wixson et al., 1987).
- Ketamine plus xylazine, combined further with acepromazine to achieve surgical anesthesia for 45 min, can have substantial cardiovascular effects, manifested by low pulse rates and hypotension (Buitrago et al., 2008).
- Ketamine plus xylazine offers sedation but does not routinely enable the animal to reach a surgical plane of anesthesia (Arras et al., 2001).
- Addition of acepromazine in the anesthetic regimen with ketamine can prolong recovery times, as determined by righting reflex and time to walking (Baker et al., 2011).

Rats

- It has been documented that rats anesthetized with ketamine plus xylazine may develop ocular lesions, including keratoconjunctivitis sicca (Kufoy et al., 1989) and irreversible corneal lesions, despite perioperative eye lubrication (Turner and Albassam, 2005).

- Profound reductions in rectal and core body temperatures have been noted in rats, demonstrated by a decrease of up to 4°C over 60 min of anesthesia (Wixson et al., 1987). This side effect emphasizes the overwhelming need to minimize hypothermia in rodents undergoing anesthesia (Lin et al., 1978, Simpson, 1997).

Ketoprofen

Rats

- In 2- to 3-month-old female Crl:CD[SD] rats, perioperative treatment with ketoprofen (5 mg/kg SC) led to marked gastrointestinal bleeding, erosions, and small intestinal ulcers, which worsened in intensity of clinical signs if the drug was coupled with inhalant isoflurane anesthesia (Shientag et al., 2012).

Oxymorphone

- Oxymorphone has been shown to be a superior analgesic for visceral pain management over buprenorphine in rats (Gillingham et al., 2001).

Tramadol

- Tramadol is an approved, opioid-like analgesic; the optimum dosage and route of administration were determined to be 12.5 mg/kg IP for provision of long-lasting and effective analgesia (Zerge Cannon et al., 2009).
- In comparative studies, tramadol (5 mg/kg SC dosed twice daily on the day of surgery and 24 h after surgery, then SID through day 3 postoperatively) has provided superior pain relief in rat models of endometriosis, as compared to buprenorphine (Debrue, 2011).
- For incisional models of pain, tramadol alone (at 10 mg/kg prior to skin incision and 10 mg/kg IP twice daily) does not provide sufficient analgesia; instead, buprenorphine (0.05 mg/kg SC) and tramadol plus gabapentin (80 mg/kg) were deemed to be appropriate (when administered preemptively and for 2 days postoperatively) (McKeon et al., 2011).

references

Arras, M, Autenried, P, Rettich, A, Spaeni, D, and Rulicke, T. 2001. Optimization of intraperitoneal injection anesthesia in mice: drugs, dosages, adverse effects, and anesthesia depth. *Comp Med* 51:443–456.

Baker, NJ, Schofield, JC, Caswell, MD, and McLellan, AD. 2011. Effects of early atipamezole reversal of medetomidine-ketamine anesthesia in mice. *J Am Assoc Lab Anim Sci* 50:916–920.

Bauer, DJ, Christenson, TJ, Clark, KR, Powell, SK, and Swain, RA. 2003. Acetaminophen as a postsurgical analgesic in rats: a practical solution to neophobia. *Contemp Top Lab Anim Sci* 42:20–25.

Bender, HM. 1998. Pica behavior associated with buprenorphine administration in the rat. *Lab Anim Sci* 48:5.

Bosgraaf, CA, Johnston, NA, and Toth, LA. 2006. Oral ibuprofen as an analgesic after abdominal surgery in mice. *J Am Assoc Lab Anim Sci* 45:117.

Bosgraaf, CA, Suchy, H, Harrison, C, and Toth, LA. 2004. What's your diagnosis? Respiratory distress in rats. *Lab Anim (NY)* 33:21–22.

Brennan, MP, Sinusas, AJ, Horvath, TL, Collins, JG, and Harding, MJ. 2009. Correlation between body weight changes and postoperative pain in rats treated with meloxicam or buprenorphine. *Lab Anim (NY)* 38:87–93.

Buitrago, S, Martin, TE, Tetens-Woodring, J, Belicha-Villanueva, A, and Wilding, GE. 2008. Safety and efficacy of various combinations of injectable anesthetics in BALB/c mice. *J Am Assoc Lab Anim Sci* 47:11–17.

Carbone, ET, Lindstrom, KE, Diep, S, and Carbone, L. 2012. Duration of action of sustained-release buprenorphine in 2 strains of mice. *J Am Assoc Lab Anim Sci* 51:815–819.

Clark, JA, Jr, Myers, PH, Goelz, MF, Thigpen, JE, and Forsythe, DB. 1997. Pica behavior associated with buprenorphine administration in the rat. *Lab Anim Sci* 47:300–303.

Clark, JA, Myers, PH, Demianenko, TK, Windham, AK, Blankenship, TL, Grant, MF, and Forsythe, DB. 2002. Analgesic potential of carprofen in mice. *Contemp Top Lab Anim Sci* 41:106.

Cowan, A, Doxey, JC, and Harry, EJ. 1977. The animal pharmacology of buprenorphine, an oripavine analgesic agent. *Br J Pharmacol* 60:537–545.

Danneman, PJ, Suckow, MA, and Brayton, CF. 2012. *The Laboratory Mouse*, 2nd edition. CRC Press, Boca Raton, FL.

Debrue, MC. 2011. Use of tramadol compared with buprenorphine to refine the postoperative care of surgically prepared endometriosis-telemetred rat models. *J Am Assoc Lab Anim Sci* 50:735–736.

Ezell, PC, Luis, P, and Lawson, GW. 2012. Palatability and treatment efficacy of various ibuprofen formulations in C57BL/6 mice with ulcerative dermatitis. *J Am Assoc Lab Anim Sci* 51:609–615.

Flecknell, PA. 2001. Analgesia of small mammals. *Vet Clin North Am Exot Anim Pract* 4:47–56, vi.

Flecknell, PA, Orr, HE, Roughan, JV, and Stewart, R. 1999. Comparison of the effects of oral or subcutaneous carprofen or ketoprofen in rats undergoing laparotomy. *Vet Rec* 144:65–67.

Flecknell, PA, Roughan, JV, and Stewart, R. 1999. Use of oral buprenorphine ("buprenorphine Jello") for postoperative analgesia in rats—a clinical trial. *Lab Anim* 33:169–174.

Foley, PL, Liang, H, and Crichlow, AR. 2011. Evaluation of a sustained-release formulation of buprenorphine for analgesia in rats. *J Am Assoc Lab Anim Sci* 50:198–204.

Gades, NM, Danneman, PJ, Wixson, SK, and Tolley, EA. 2000. The magnitude and duration of the analgesic effect of morphine, butorphanol, and buprenorphine in rats and mice. *Contemp Top Lab Anim Sci* 39:8–13.

Gaertner, DJ, Hankenson, FC, Hallman, T, and Batchelder, MA. 2008. Anesthesia and analgesia in rodents, Chap. 10. In Fish, RE, Brown, MJ, Danneman, PJ, and Karaz, AZ (eds.), *Anesthesia and Analgesia for Laboratory Animals*. Academic Press, San Diego, CA.

Gillingham, MB, Clark, MD, Dahly, EM, Krugner-Higby, LA, and Ney, DM. 2001. A comparison of two opioid analgesics for relief of visceral pain induced by intestinal resection in rats. *Contemp Top Lab Anim Sci* 40:21–26.

Goecke, JC, Awad, H, Lawson, JC, and Boivin, GP. 2005. Evaluating postoperative analgesics in mice using telemetry. *Comp Med* 55:37–44.

Hackett, TB, and Lehman, TL. 2005. Practical considerations in emergency drug therapy. *Vet Clin North Am Small Anim Pract* 35:517–525, viii.

Hausamann, JC, Anderson, LC, Mariana, M, Ronan, J, Conarello, S, Capodanno, I, Nunes, C, and Fenyk-Melody, JE. 2002. Acetaminophen-induced hepatocellular necrosis: a comparison in male Sprague Dawley, Wistar and Lewis rats. *Contemp Top Lab Anim Sci* 41:113–114.

He, S, Atkinson, C, Qiao, F, Chen, X, and Tomlinson, S. 2010. Ketamine-xylazine-acepromazine compared with isoflurane for anesthesia during liver transplantation in rodents. *J Am Assoc Lab Anim Sci* 49:45–51.

IACUC-UPENN. 2010. IACUC Guideline. Rodent Anesthesia and Analgesia Formulary. http://www.upenn.edu/regulatoryaffairs/Documents/iacucguideline-rodentanesthesiaandanalgesiaformulary.pdf

Interagency Research Animal Committee (IRAC). 1985. *U.S. Government Principles for Utilization and Care of Vertebrate Animals Used in Testing, Research, and Training.* Interagency Research Animal Committee; Office of Science and Technology Policy, Washington, DC.

Kufoy, EA, Pakalnis, VA, Parks, CD, Wells, A, Yang, CH, and Fox, A. 1989. Keratoconjunctivitis sicca with associated secondary uveitis elicited in rats after systemic xylazine/ketamine anesthesia. *Exp Eye Res* 49:861–871.

Libbin, RM, and Person, P. 1979. Neonatal rat surgery: avoiding maternal cannibalism. *Science* 206:66.

Lin, MT, Chen, CF, and Pang, IH. 1978. Effect of ketamine on thermoregulation in rats. *Can J Physiol Pharmacol* 56:963–967.

Martin, LB, Thompson, AC, Martin, T, and Kristal, MB. 2001. Analgesic efficacy of orally administered buprenorphine in rats. *Comp Med* 51:43–48.

McKeon, GP, Pacharinsak, C, Long, CT, Howard, AM, Jampachaisri, K, Yeomans, DC, and Felt, SA. 2011. Analgesic effects of tramadol, tramadol-gabapentin, and buprenorphine in an incisional model of pain in rats (*Rattus norvegicus*). *J Am Assoc Lab Anim Sci* 50:192–197.

Meador, CK. 1998. Polypharmacy: old bad habits. *J Am Board Fam Pract* 11:166–167.

Molloy, MD, Garcia, FG, and Olin, JM. 2004. The use of isoflurane and hypothermia as an anesthetic protocol for surgical subcutaneous implants in neonatal rats. *Contemp Top Lab Anim Sci* 43:82.

National Research Council (NRC). 2011. *Guide for the Care and Use of Laboratory Animals*, 8th edition. National Academies Press, Washington, DC.

Oglesbee, BL. 2011. Rodents, pp. 544–622. In Oglesbee, BL (ed.), *Blackwell's Five-Minute Veterinary Consult: Small Mammal*, 2nd edition. Wiley-Blackwell, Ames, IA.

Ovadia, S, and Zeiss, CJ. 2002. Ptyalism and anorexia in a Sprague-Dawley rat. *Lab Anim (NY)* 31:25–27.

Parker, JM, Austin, J, Wilkerson, J, and Carbone, L. 2011. Effects of multimodal analgesia on the success of mouse embryo transfer surgery. *J Am Assoc Lab Anim Sci* 50:466–470.

Phifer, CB, and Terry, LM. 1986. Use of hypothermia for general anesthesia in preweanling rodents. *Physiol Behav* 38:887–890.

Plumb, DC. 2005. *Plumb's Veterinary Drug Handbook*. Wiley-Blackwell, Ames, IA.

Shientag, LJ, Wheeler, SM, Garlick, DS, and Maranda, LS. 2012. A therapeutic dose of ketoprofen causes acute gastrointestinal bleeding, erosions, and ulcers in rats. *J Am Assoc Lab Anim Sci* 51:832–841.

Simpson, DP. 1997. Prolonged (12 hours) intravenous anesthesia in the rat. *Lab Anim Sci* 47:519–523.

Speth, RC, Smith, MS, and Brogan, RS. 2001. Regarding the inadvisability of administering postoperative analgesics in the drinking water of rats (*Rattus norvegicus*). *Contemp Top Lab Anim Sci* 40:15–17.

Takeda, N, Hasegawa, S, Morita, M, and Matsunaga, T. 1993. Pica in rats is analogous to emesis: an animal model in emesis research. *Pharmacol Biochem Behav* 45:817–821.

Thompson, AC, Kristal, MB, Sallaj, A, Acheson, A, Martin, LB, and Martin, T. 2004. Analgesic efficacy of orally administered buprenorphine in rats: methodologic considerations. *Comp Med* 54:293–300.

Turner, PV, and Albassam, MA. 2005. Susceptibility of rats to corneal lesions after injectable anesthesia. *Comp Med* 55:175–182.

Weiner, CM, Multari, H, Wilwol, M, Boutin, S, and Lohmiller, JJ. 2011. A nonsteroidal antiinflammatory drug improves surgical outcome in hypophysectomized animals. *J Am Assoc Lab Anim Sci* 50:747.

Wixson, SK, and Smiler, K.L. 1997. Anesthesia and analgesia in rodents, pp. 166–203. In Kohn, DF, Wixson, SK, White, WJ, and Benson, GJ (eds.), *Anesthesia and Analgesia of Laboratory Animals*. Academic Press, New York.

Wixson, SK, White, WJ, Hughes, HC, Jr, Lang, CM, and Marshall, WK. 1987. The effects of pentobarbital, fentanyl-droperidol, ketamine-xylazine and ketamine-diazepam on core and surface body temperature regulation in adult male rats. *Lab Anim Sci* 37:743–749.

Yamada, K, Popilskis, SJ, and Lipman, NS. 2009. Assessment of the efficacy and duration of buprenorphine and carprofen analgesia in a mouse model of acute incisional pain. *J Am Assoc Lab Anim Sci* 48:79.

Zerge Cannon, C, Waxer, D, Myers, P, Clark, J, Goulding, DR, King-Herbert, AP, and Blankenship, T. 2009. Evaluation of route of administration and dosage of tramadol as an analgesic in the rat. *J Am Assoc Lab Anim Sci* 48:91.

Index

a

b

j

k

l

m

n

P

q

r

S

t